Energy Issues

New Directions and Goals

Timely Reports to Keep
Journalists, Scholars and the Public
Abreast of Developing Issues, Events and Trends

Editorial Research Reports
Published by Congressional Quarterly Inc.
1414 22nd Street, N.W.
Washington, D.C. 20037

About the Cover

Cover illustration by Guy Aceto; designed by Art Director Richard Pottern.

Editor, Hoyt Gimlin
Managing Editor, Sandra Stencel
Editorial Assistants, Laurie De Maris, Nancy Blanpied

Production Manager, I. D. Fuller
Assistant Production Manager, Maceo Mayo

Library Congress Cataloging in Publication Data
Main entry under title:

Editorial research reports on energy issues.

    Bibliography: p.
    Includes index.
    1. Power resources.   2. Energy policy.   I. Congressional Quarterly, inc.
TJ163.2.E33   333.79   82-2523
ISBN 0-87187-234-X   AACR2

Contents

Foreword

When Editorial Research Reports published the book *Earth, Energy and Environment* in January 1977, the country was in the midst of an energy crisis. A severe cold spell in the East and Midwest caused such rapid depletion of the nation's declining natural gas supplies that many schools and factories were closed. In a more than symbolic gesture, President Carter ordered White House thermostats turned down to a chilly 65 degrees and asked all Americans to follow his example.

Just five years later, much had changed. The winter of 1981-82 will be remembered in many areas for its blizzards and record cold temperatures. But while the snow and bone-chilling cold taxed the psychological resources of most Americans, the winter passed without significant energy shortages. As spring arrived, oil glutted the market, forcing gasoline prices to their lowest levels in years.

Most experts believed the excess oil supplies were temporary — the product of conservation, a deep recession and disagreements over production levels among the members of the Organization of Petroleum Exporting Countries.

During his first year in office, President Reagan gave energy policy a lower priority than his predecessors. The administration's emphasis was on energy production. Funding for conservation programs, solar energy and synthetic fuel development was cut back, while spending on nuclear energy increased.

Reagan's proposed fiscal 1983 budget included a plan to dismantle the Department of Energy, fulfilling a longstanding Reagan promise. The plan faced strong opposition in Congress. The administration also sought to push ahead with deep-sea mining, undercutting a decade of U.N.-sponsored efforts to negotiate a worldwide Law of the Sea treaty.

Despite shrinking federal subsidies, public interest in alternative energy technologies remained high. Americans spent millions of dollars on wood stoves, windmills, solar water heaters and homes built with passive solar features. There also was much skepticism about the administration's support for nuclear energy. Public concern centered on the nation's growing backlog of nuclear waste and the lack of a national policy for dealing with the problem.

The 10 reports included in this book discuss these and other issues that could affect the direction of the nation's energy policy.

Sandra Stencel
Managing Editor

May 1982
Washington, D.C.

ENERGY POLICY: THE NEW ADMINISTRATION

by

William Sweet

Jan. 30
1 9 8 1

Editor's Note: During his first year in office, President Reagan gave energy policy a lower priority than his predecessors. While the measures the president favored were generally consistent with the free-market approach he advocated, Reagan did not always push this approach as hard or as far as he might have. The new administration cut federal funding for conservation, solar energy, fossil fuels and synthetic fuel development; boosted funding for nuclear energy, including the controversial Clinch River fast breeder reactor; and attempted to relax government restrictions on energy resource development. The administration did not abolish the U.S. Synthetic Fuels Corp., however, and it did not try to accelerate deregulation of natural gas prices.

The administration's proposed budget for fiscal 1983 would abolish the Department of Energy and transfer most of the department's programs to a newly created division of the Commerce Department. But the plan has run into sharp opposition among both Democrats and Republicans in Congress. In March 1982, Reagan vetoed a bill that would have given him the authority to allocate energy supplies during a national crisis.

ENERGY POLICY:
THE NEW ADMINISTRATION

SINCE 1973, when world oil prices took their first giant leap upward, Americans have heard from all sides that they must learn to adjust to an age of scarcity. During his four-year term, President Carter repeatedly told Americans that they would have to use less energy and pay more for it in order to be sure of having adequate supplies for the future. Responding to such admonitions and to rising energy prices, Americans did begin to conserve. Motorists switched to more fuel-efficient cars; homeowners and managers of commercial property installed insulation and improved heating systems; businesses started to develop and use energy-saving industrial processes.

In sharp contrast to Carter's call for collective sacrifice, President Reagan has made it clear that his administration will emphasize energy production rather than conservation. Echoing this view, Energy Secretary James B. Edwards said recently that "most Americans now agree we must increase production of our own energy resources." Testifying Jan. 12 at his confirmation hearings before the Senate Committee on Energy and Natural Resources, the former South Carolina governor argued that greater production by the private sector — not conservation — is the key to a vital and alert America. Comparing the country to a person, Edwards said: "The human body uses least energy when it's asleep or dead."

In his acceptance speech at the Republican convention in July, Reagan condemned "those who preside over the worst energy shortage in our history" for telling us "to use less, so that we will run out of oil, gasoline and natural gas a little more slowly." Reagan said that the United States "must get to work producing more energy. . . . Large amounts of oil and natural gas lie beneath our land and off our shores. . . . Coal offers great potential. So does nuclear energy produced under rigorous safety standards. . . . It must not be thwarted by a tiny minority opposed to economic growth which often finds friendly ears in regulatory agencies for its obstructionist campaigns."

The platform that the Republicans adopted at their convention in Detroit endorsed a "strategy of aggressively boosting the nation's energy supplies." To that end, the platform called for (1) decontrol of oil and gas prices "at the wellhead"; (2)

repeal of the "so-called windfall profits tax" on domestic oil, which Congress enacted in 1980 at the behest of President Carter; (3) enactment of a "comprehensive program of regulatory reform" and "revision of overly stringent Clean Air Act regulations"; (4) upgrading of the nation's coal transportation systems and giving state governments the primary responsibility for enforcement of strip-mining regulations; (5) "accelerated use of nuclear energy"; and (6) opening of federal lands and the Outer Continental Shelf for energy development.

The platform did not take a position on the controversial Synthetic Fuels Corporation, which Congress set up last year to develop technologies for conversion of coal and oil shale to oil and gas fuels *(see p. 11)*. While the platform criticized the Department of Energy, which the Carter administration created in 1977, it did not call directly for the department's abolition. During the campaign, however, Reagan said that both the Energy Department and the Synthetic Fuels Corporation should be eliminated.

Energy Department's Future Under Reagan

During the weeks following the presidential election, when Reagan was filling Cabinet positions, it was well known that the energy post was "finding few seekers."[1] When Reagan announced his selection of Edwards in late December, his choice of a man who seemed to have a weak background in energy policy was taken as a bad omen for the department *(see profile, p. 5)*.[2] A dental surgeon by profession, Edwards said at the time: "I'd like to go to Washington and close the Energy Department down and work myself out of a job." But even before his confirmation hearings began, Edwards was talking about "streamlining" the department rather than shutting it down. And in the hearings, under pressure from ranking committee members who had helped create the department, Edwards said he was "rethinking" the idea of abolishing the department, though he still favored "dismantling some of the regulatory apparatus."

During the first two years of the Energy Department's existence, when James R. Schlesinger was in charge, the department often was characterized as an organizational mess. Schlesinger was said to be interested primarily in policy, not administration, and the organizational ideas he had were thought to be unworkable by many people in the department. During the gasoline shortage of early 1979, which took place at a time when Iranian oil production had dropped sharply as a result of the country's revolution, the manner in which

---

[1] See *The Wall Street Journal*, Dec. 15, 1980.
[2] It was reported that Edwards had been selected primarily because Sen. Strom Thurmond, R-S.C., had pressured Reagan to include a Southerner in the Cabinet.

James B. Edwards

A successful dental surgeon, Edwards was governor of South Carolina from 1974 to 1978. As governor, he took the lead in establishing the South Carolina Energy Research Institute to evaluate the state's energy options. He served as chairman of the National Governors Association's nuclear subcommittee, and he is known as a strong supporter of nuclear power, a major industry in South Carolina.

During Senate confirmation hearings, Edwards was able to handle detailed questions about energy policy without getting

himself in trouble. He endorsed accelerated decontrol of oil and gas prices, more federal support for nuclear energy, more rapid filling of the strategic petroleum reserve, opening of federal lands to energy development and streamlining of permit procedures for energy projects.

Committee members raised a number of questions about Edwards' sensitivity to environmental concerns. In 1974, Edwards was a member of a partnership that developed a South

James B. Edwards

Carolina island over objections from the U.S. Corps of Engineers, causing — according to the Corps — "irreparable damage" to the island's ecology. In 1975, Edwards was reported to have said pollution control laws were "not needed here in South Carolina where we have nice breezes to carry off the emissions and dissipate them."

Schlesinger managed allocation of fuel supplies came under especially sharp attack.[3] At least in part because of Schlesinger's poor public image, Carter replaced him in mid-1979 with Charles W. Duncan, a former president of Coca-Cola Co.

Duncan managed to get the department out of the daily headlines, and his personality was widely perceived as less abrasive than Schlesinger's. But opinions differ as to whether administration of the department improved significantly during his tenure. The new energy secretary, Edwards, has said he will evaluate the department and streamline it. But Edwin Meese III, a top Reagan adviser with Cabinet rank, told reporters Jan. 16 — after Edwards' confirmation hearing had ended — that Reagan will abolish the Energy Department just as soon as the secretary can tell him how to do it.

Diverse Influences on New Energy Policy

Edwards made it clear during his confirmation hearings that he expects to be the administration's principal spokesman on energy and the "initiator" of policy. Energy policy has such

[3] See "Public Confidence and Energy," *E.R.R.*, 1979 Vol. I, pp. 385-389.

wide ramifications, however, that many analysts expect the
White House staff and various Cabinet-level officials to have
an important and perhaps even dominating influence on the
formulation of key administration positions. Reagan's adviser
for domestic affairs, Martin Anderson, could end up a kind
of energy policy coordinator, and at least three other high officials will have considerable influence.

---

*"Most Americans now agree we must in-
crease production of our energy resources."*

Secretary of Energy James B. Edwards
Testifying before the Senate Energy
Committee, Jan. 12, 1981

One, the administrator of the Environmental Protection
Agency, has yet to be appointed. But Reagan is expected to
pick somebody who will be willing to set, interpret and enforce
standards in a manner that is conducive to greater energy
production. *The Washington Post* reported Jan. 21 that Reagan
planned to name James R. Mahoney, a meteorologist and business consultant, to the EPA post. Mahoney is co-founder and
senior vice president of Environmental Research and Technology Inc., an industry-oriented consulting and environmental
monitoring company based in Concord, Mass. He is considered
a political conservative and an expert on clean air legislation.
Whoever is appointed EPA administrator will play a key role
in recommending and commenting on proposed revisions of
the 1970 Clean Air Act and its 1977 amendments. Many industrial leaders regard relaxation of some clean air rules as crucial
to the viability of new energy projects, especially the coal,
oil and oil-shale development programs slated for the Western
states.[4]

Another official who will have a lot to say about energy
development is James G. Watt, the new secretary of interior.
During his confirmation hearings, Watt said Reagan had outlined four key goals for the Interior Department, the first of
which was to make public lands available for more uses, including oil exploration and coal mining. Gaylord Nelson, the former
Democratic senator from Wisconsin who now heads the Wilder-

---

[4] See "Air Pollution Control," *E.R.R.*, 1980 Vol. II, pp. 841-864. Even before Reagan
took office, his transition team tried without success to get EPA to delay issuing new
water pollution rules for coal mines and certain coal-processing plants.

ness Society, testified that Watt "brings to his office what we believe to be a strong anti-environmental bias."[5]

EPA and Interior Department decisions relating to energy policy are virtually certain to unleash stormy battles in the coming years. But the official who may well end up at the eye of the biggest hurricane is David A. Stockman, the new head of the White House Office of Management and Budget (OMB). As U.S. representative from Michigan's fourth district, Stockman developed a reputation for being one of the Republican Party's smartest and hardest-driving young members. Just 34-years-old, Stockman developed considerable expertise on environmental policy as a member of the House Commerce Committee and the National Commission on Air Quality. As a doctrinaire free-market conservative, Stockman has strongly opposed federal subsidies for energy projects, including government spending on advanced nuclear technologies and synthetic fuels.

Divisions Over Spending and Environment

Orthodox free-marketeers like Stockman often disagree sharply with Republicans who have the interests of the big energy industries at heart. Because of such divisions, peculiar coalitions of Republicans and Democrats could emerge, similar to the one that arose when Carter proposed an "Energy Mobilization Board" to expedite energy projects. The House unexpectedly rejected the board last year, partly because Western conservatives regarded it as an intrusion on states' rights, and partly because liberals feared it would lead to neglect of environmental, health and safety standards.[6]

At his confirmation hearing, Energy Secretary Edwards indicated that he supported breeder reactor development and reprocessing of spent nuclear reactor fuels. Stockman is known to oppose these projects. A coalition of like-minded fiscal conservatives and members of Congress concerned about arms control — including some Republicans — could end up prevailing on these issues *(see box, p. 19)*. Edwards also supports continued government funding for development, but not commercialization, of synthetic fuels — a program Stockman has vigorously opposed. Environmentalists also oppose the synfuels program, and they could join with fiscal conservatives to halt or curtail its development.

---

[5] As former president of the Mountain States Legal Foundation based in Denver, Watt was involved in numerous suits against the Interior Department, many of which aimed to open public lands to development by mining, oil and timber interests. Watt supported the "Sagebrush Rebellion," a movement to transfer control over millions of acres of federally owned land to the Western states. See "Rocky Mountain West: An Unfinished Country," *E.R.R.,* 1980 Vol. I, pp. 185-204.

[6] See *Congressional Quarterly Weekly Report,* June 28, 1980, pp. 1790-1792.

The overall balance of forces in the 97th Congress has yet to be tested, and the net effect of the many important committee changes has yet to be gauged. Under the leadership of Sen. James A. McClure, R-Idaho, the Energy and Natural Resources Committee probably will be sympathetic to synfuels,the nuclear industry and the big energy corporations in general; under Democratic leadership the committee's positions were about the same. Sen. Robert T. Stafford, R-Vt., will be a strong spokesman for the environment as the new chairman of the Committee on the Environment and Public Works, though he may not be as effective as his predecessor, former Democratic Sen. Edmund S. Muskie of Maine. On the House side, Rep. Morris K. Udall, D-Ariz., will continue to be an important environmental advocate as chairman of the Interior Committee. Leaders of the Commerce Committee are expected to fight to retain government support for conservation measures and "alternative" energy technologies — solar energy, geothermal steam, etc.[7]

It is not at all clear how high a priority the Reagan administration will assign energy policy, but it seems probable that energy will not be given as much emphasis as it got under Carter. On many controversial issues, Reagan may prefer to stall or compromise rather than stake a lot of political capital on issues that are highly divisive even within the Republican Party. Reagan has moved quickly, on the other hand, in the one area where virtually all Republicans agree — accelerated decontrol of energy prices. The White House announced Jan. 28 the immediate termination of controls on oil, which were due to expire on Sept. 30 under Carter's program. He also is expected to endorse terminating controls on natural gas, which are to be phased out by 1985. But accelerated decontrol of gas prices will require legislation, and a legislative battle could develop over this issue.

Carter's Energy Legacy

ESTIMATES differ as to how great the impact of immediate oil decontrol will be on prices for gasoline and home heating oil. Roughly three-fourths of the oil produced in the

---

[7] Rep. John D. Dingell, D-Mich., is the new chairman of the full committee. The Subcommittee on Energy and Power, which Dingell used to direct, was scheduled to be split into two subcommittees on Jan. 28, one on fossil fuels and one on nuclear energy, utilities, conservation and alternative energy technologies. Rep. Philip R. Sharp, D-Ind., will lead the fossils panel, and Rep. Richard L. Ottinger, D-N.Y., will head the other subcommittee.

The Frustrating Search for More Oil

	U.S. Oil Reserves (billion barrels)	U.S. Gas Reserves (trillions of cubic feet)	Oil and Gas Wells Drilled (thousands)
1970	39.0	290.8	28.1
1971	38.1	278.8	25.9
1972	36.3	266.1	27.3
1973	35.3	250.0	26.6
1974	34.3	237.1	31.7
1975	32.7	228.2	39.1
1976	31.0	216.0	41.5
1977	29.5	208.9	46.5
1978	27.8	200.3	48.5
1979	27.1	194.9	51.2
1980	—	—	62.7*

Source: American Petroleum Institute.

* Source: Petroleum Information Service.

United States already had been decontrolled by the time Reagan took office, and some of the remaining oil that will be affected by the Reagan order is scheduled for purchase by the government as part of the program to fill a strategic oil reserve (see p. 20). Albert Linden, acting administrator of the Energy Information Administration, told the Senate Energy Committee on Jan. 22 that immediate decontrol might lead to a rise in gasoline and home heating oil prices of 8 to 10 cents a gallon. Stockman has said prices might rise by about 12 cents a gallon.

The basic work of decontrolling oil prices was carried out by President Carter between April 1979, when he announced he would lift controls, and when he left office. To many people, it may seem ironic that the one and only energy policy on which virtually all Republicans can agree — price decontrol — is the one policy which already had been introduced by a Democratic Congress and a Democratic president. But in fact, the price decontrol policies that emerged as the heart of Carter's energy policy were not originally part of his program, and they were not the policies that he advocated during the 1976 campaign.

In the spring of 1977, Carter proposed a sweeping energy program. The plan included incentives for property owners to insulate their buildings, mandatory energy efficiency standards for industrial products, prohibitions on industrial use of certain fuels, a stiff tax on "gas guzzling" cars and a "wellhead" tax on domestic oil — that is, when it is pumped out of the ground. The program's emphasis on conservation appealed to environmentalists, and its heavy reliance on mandatory procedures rather than high prices impressed equity-minded Democrats as a fair approach to the conservation problem.

The House of Representatives, which was much more liberal on economic issues than the Senate was during the Carter years, passed much of Carter's energy program with dispatch. The program got bogged down in the Senate, however, and most of what Carter had proposed ended up getting passed only in a watered-down version, if at all. The National Energy Conservation Policy Act of 1978 authorized mandatory efficiency standards for some home appliances and required utilities to provide customers with information about conservation devices and procedures. The Public Utility Regulatory Policies Act, also passed that year, required state regulatory commissions to "consider" setting rate structures that reflected the actual (rather than the average) costs of supplying power. The Powerplant and Industrial Fuel Use Act of 1978, following up on legislation enacted four years earlier, prohibited new utility plants from burning oil or gas and required existing plants to phase out use of those fuels by 1990.[8]

Oil and Gas Decontrol; Synfuel Proposals

Of all the energy policies debated in 1977 and 1978, natural gas pricing provoked the bitterest battles. Carter had made contradictory promises about gas pricing during the 1976 campaign, and Congress was sharply divided on the issue. Virtually all Republicans favored letting the market set gas prices, but liberals representing the gas-consuming states of the Northeast and Midwest tended to be strongly against decontrol. Environmentalists were divided between those who favored letting prices rise as a stimulus to conservation, and those who thought conservation should be based on rationing and other mandatory measures. After a protracted battle, Congress decided in 1978 to phase out controls on natural gas by 1985.

At the beginning of 1979, when world oil prices began to rise very rapidly for the first time since the 1973-74 Arab oil embargo, Carter officials could take little satisfaction in what they had achieved toward establishing an energy policy. Congress had rejected or weakened most of what Carter had proposed in 1977. It was under these circumstances, with the country facing rising gasoline prices and occasional gasoline shortages, that Carter made a surprising new departure.

---

[8] The Energy Supply and Environmental Coordination Act of 1974 had ordered utilities to stop using oil and gas. The implementation of this law turned out to cause serious environmental problems, and many utilities switched back to petroleum after a brief flirtation with coal.

On April 5, 1979, Carter announced he would phase out controls on domestic oil prices, acting under authority provided by the Energy Policy and Conservation Act of 1975. To prevent U.S. oil companies from reaping "huge and undeserved profits" as a result of decontrol, Carter asked Congress to impose a "windfall profits tax" on the added revenues that oil companies would obtain. Revenues from the windfall profits tax, as Carter proposed it at that time, were to go for mass transit, programs to reduce U.S. reliance on foreign oil and an Energy Security Trust Fund to help low-income families deal with higher energy costs.

"Without exaggeration ... conservation was a 'source' of energy during the 1970s [and will be] a significant source in the future, also."

The Brookings Institution
Setting National Priorities (1980)

Carter apparently hoped to whip up strong public support for the windfall profits tax, but the public did not rush to his side, and Congress was slow in acting. In mid-summer 1979, when polls were indicating that confidence in America's future had fallen to a record low, Carter went to Camp David and conferred with a large number of leading Americans for 10 days. Upon returning to Washington from the mountain retreat, on July 15, Carter delivered a nationally televised speech in which he proposed an $88 billion, 10-year program to boost production of synthetic oil and gas, to be funded by revenues from the windfall profits tax.[9]

In March 1980, Congress finally passed a windfall profits tax on oil, designed to raise roughly $227 billion over a decade out of about $1 trillion in added oil company revenues. But Congress rejected Carter's idea of putting the revenues into a special trust fund and instead funneled the money into general revenues. The bill did specifically authorize tax incentives to businesses and homeowners for conservation, and it provided $3.1 billion in block grants to the states for fuel assistance to the poor.

The following June, Congress passed the Energy Security Act, which set up a Synthetic Fuels Corporation with a $20

---

[9] See "Synthetic Fuels," *E.R.R.*, 1979 Vol. II, pp. 62-640.

billion authorization. The corporation is to submit a comprehensive plan for synfuels development by 1984, and at that time it may be eligible for another $68 billion in federal funds. In addition to providing funds for production of fuels from coal, oil shale and alcohol ("gasohol"), the bill authorized loans and grants for solar energy and conservation measures.

Many energy experts and economists regard synthetic fuels as an excessively expensive program, especially since it is not expected to free the United States completely from dependence on foreign oil.[10] Only two countries have made large public expenditures on synfuels, Germany under the Nazis and South Africa today. The Nazis considered the expense worthwhile because they could not fight a big war without having a reliable domestic supply of oil. South Africa spends the money because it fears it may be made the target of an oil embargo because of its oppressive racial policies. To many observers, the energy problems facing the United States — however serious — appear much less drastic than those which have justified synfuels development in other cases.

Changes Foreseen By New Administration

Despite Stockman's opposition to synfuels, the conventional wisdom has it that the Republicans will retain the synfuels program, if only because "it contains too much pork for the Republicans to throw away."[11] Many Western states, where Republicans are politically powerful, would benefit economically from the synfuels program. But if the Republicans find it even harder than expected to cut federal spending on social programs while increasing expenditures on defense at the same time, synfuels could fall prey to the budget cutters.

In their zeal to get federal spending down, the Republicans are likely to reduce outlays for many of the smaller energy programs set up during the past four years. Funding for alternative energy technologies could be affected. So could some of the tax incentives provided for conservation measures. But wholesale abolition of such programs is considered unlikely. Most have some support in both parties, and most just aren't big enough to have a dramatic impact on the budget deficit.

If the Republicans do make a major legislative initiative on energy in 1981, it is likely to be on natural gas pricing. There is strong support in the party for accelerating the process of decontrol, which producers of natural gas strongly favor.

---

[10] The United States imported about 8 million barrels of oil a day in 1979. The synfuels program is intended to provide the equivalent of 2 million barrels a day by 1992.

[11] An anonymous Democratic staff member of the House Energy Committee, quoted by Eliot Marshall in "An Early Test of Reagan's Economics," *Science,* Jan. 2, 1981, p. 30.

The advocates of price decontrol argue that exploration for natural gas already has picked up nicely in response to the phased decontrol of prices, and they conclude that more rapid decontrol would produce even more intense exploration and production.[12] Distributors of natural gas are opposed to decontrol, on the other hand, because they assume that higher prices will compress demand. Only time will tell whether the Reublicans choose to grapple with this contentious issue in 1981.[13]

Conservation and Production

ONE of the main reasons why price decontrol has emerged as the key part of the nation's energy policy, and certainly the reason why it has won grudging support from some people who initially opposed it, is that higher prices are supposed to have favorable effects — from the point of view of energy independence — on both production and consumption. With millions of consumers and thousands of producers making decisions about energy every day, only price decontrol seems to send the right signal to everybody simultaneously. Even so, questions have been raised about whether decontrol really is as effective as it sounds. Consumers will conserve in response to higher prices, critics point out, only if there are viable and economic means of conserving. And more energy will be produced in response to higher prices only if there is more energy out there to be found.

On the consumption side, the results of the last few years have been encouraging. According to the Brookings Institution's 1980 overview of the energy situation, the U.S. economy as a whole achieved a 10 percent improvement in aggregate energy efficiency between 1972 and 1978. "Without exaggeration...," the Brookings study observed, "conservation was a 'source' of energy during the 1970s. Analysts expect conservation to be a significant source in the future, also. Projections for U.S. energy demand in the year 2000, which ranged between 150 quadrillion and 175 quadrillion British thermal units (Btu) a

---

[12] Edward W. Erickson, consultant to the Natural Gas Supply Association, estimates that the final statistics for 1980 will show that 15 trillion cubic feet of natural gas were discovered, independent of oil exploration, compared to 10.4 trillion cubic feet in 1979.

[13] According to the Jan. 13, 1981, issue of *Energy Daily*, the Reagan transition team at the Energy Department wanted the administration to send Congress a comprehensive natural gas bill early in 1981. In addition to accelerating decontrol of gas prices, the transition team advocated modifying restrictions on use of natural gas by utilities and abolishing pricing rules that discriminate in favor of residential customers and against industrial users.

Oil Imports

(thousands of barrels per day)

	Net Imports*	OPEC	Non-OPEC
1973	6,025	2,992.9	3,263.2
1974	5,892	3,279.8	2,832.4
1975	5,846	3,601.3	2,454.4
1976	7,090	5,065.8	2,246.8
1977	8,565	6,193.1	2,614.1
1978	8,002	5,750.9	2,612.5
1979	7,939	5,612.0	2,799.1

* U.S. imports of oil minus U.S. exports of oil.

Source: Department of Energy.

decade ago, now cluster around 110 quadrillion to 120 quadrillion."[14]

Several studies have concluded during the past two years that conservation — meaning the more efficient use of energy resources, without detriment to the nation's standard of living — is the most promising single means of reducing the country's dependence on imported oil.[15] As recently as Jan. 16, a panel of the President's Commission for a National Agenda for the Eighties reported that conservation is the most promising energy policy option. The panel recommended that funds now earmarked for synthetic fuel projects and proceeds from the windfall profits tax be applied to energy-saving measures.[16]

To a great extent, it would seem that the United States only needs to copy procedures already in wide use to achieve very substantial efficiency improvements. In Germany and France, per capita consumption of energy is much lower than in the United States, and 50 to 75 percent less energy is used to produce each unit of output. Yet the quality of life in those countries is similar to America's, and their products compete successfully with U.S. goods in world markets.

Conservation efforts in the United States so far have been more successful in the industrial and commercial sectors than in the residential sector. The results achieved in the automobile industry have been especially striking. Manufacturers have met

[14] Hans H. Landsberg, "Energy," in *Setting National Priorities* (1980), p. 105. U.S. consumption of energy per unit of gross domestic product dropped from 61,574 Btu in 1960 to 56,613 Btu in 1978. A British thermal unit is the amount of heat required to increase the temperature of a pound of water one degree Fahrenheit.

[15] See *Energy Future*, the Harvard Business School study edited by Robert Stobaugh and Daniel Yergin (1979), and *Energy in Transition, 1985-2000*, Committee on Nuclear and Alternative Energy Systems, National Research Council (1980).

[16] The Panel on Energy, Natural Resources and the Environment was headed by Dr. Daniel Bell, a professor of sociology at Harvard University. The commission was created by President Carter in October 1979. See "America in the 1980s," *E.R.R.*, 1979 Vol. II, pp. 861-880.

government requirements for improvements in auto fuel efficiency on schedule, and industry leaders say they would have had to make the improvements even in the absence of mandatory schedules since consumers were turning to fuel-efficient cars in droves. According to Roger B. Smith, the new chairman of General Motors Corp., GMC cars have attained a 92 percent improvement in fuel efficiency, which he describes as the best performance of any U.S. auto company.[17]

A number of reasons help explain why gains in the residential sector have been somewhat less impressive. Since there are many more residential consumers than there are industrial users, more time necessarily is required to disseminate information on conservation techniques. When renters of property pay their own utility bills, landlords have little incentive to invest in insulation or fuel-efficient equipment. And when landlords pay the bills, renters have little incentive to use energy more efficiently. Even in cases where people own their own homes, they often lack an economic incentive to make improvements that eventually would pay for themselves. Savings from conservation procedures often exceed investment costs over a 20-year or 30-year period. But many homeowners are not interested in making the investments because they don't expect to be in their dwellings that long.

Because of this problem, Congress has provided tax credits to homeowners who insulate their houses or purchase alternative energy systems such as solar heaters. In 1979, 4.8 million taxpayers took advantage of these credits, and the year before that 5.9 million did so. Some analysts believe that even more impressive results could be achieved if the government imposed mandatory conservation standards on rental property or established a sort of "conservation corps" to carry out improvements in residential property at public expense.

Outlook for Coal and the Nuclear Industry

While prospects for conservation look good across the board, the outlook for enhanced energy production is more mixed. Coal, the most promising of the domestic fuels, has attracted enormous investments in recent years, above all from the major oil companies, utilities and steel corporations. Oil and gas companies now own roughly 55 billion tons of U.S. coal reserves, about 40 percent of non-governmental tonnage. The new investments have gone largely to Western coal, which has a low sulfur content and generally can be strip-mined, unlike most of the coal found in the East and Midwest *(see box, p. 17)*.

According to preliminary estimates, U.S. coal production

---

[17] Smith made this claim in a speech to the National Press Club in Washington, D.C., on Jan. 16, 1981.

climbed to 824 million tons last year, up 7 percent from 1979 and close to 300 million tons higher than in 1960. The added production has gone not only to domestic consumers but also to the world market, in which the United States is an ever more important supplier. An international report issued May 12, 1980, predicted that the United States may well become the "Saudi Arabia" of world coal exporters.[18] At the economic summit meeting of Western leaders who convened in Venice last June, President Carter got the Europeans to agree that the West should increasingly rely on U.S. coal.

There are a number of important barriers to sharply increased use of coal in the United States: (1) distribution bottlenecks, including inadequate port facilities, a decaying railroad system and lack of water to support proposed coal-slurry pipelines in the West; (2) insufficient storage facilities; (3) clean air regulations, which place restrictions on both the production and consumption of coal; and (4) strip-mining regulations, especially the exacting standards for restoration of mined land.

Some of these barriers, for example the supposedly adverse impact of clean air rules on the conversion of power plants from oil to coal, may have been overestimated.[19] In many cases, private industry may overcome obstacles without special assistance. And in some instances — the need to improve ports, for example — Congress may agree on a federal program without much fuss.

But many of the barriers to added coal consumption continue to arouse strong passions, and efforts to dismantle or lower these barriers may provoke some of the fiercest legislative battles of the early Reagan years. The coming revision of clean air legislation is almost universally expected to involve a big fight. And there is almost sure to be a highly controversial effort to rewrite the Surface Mining and Reclamation Act of 1977, which was passed only after a six-year struggle and two presidential vetoes (by Gerald R. Ford).

The 1977 law set performance standards for environmental protection to be met at most surface mining operations for coal. It also provided for joint responsibility and enforcement by the states and the federal government, established a self-supporting Abandoned Mine Reclamation Fund to restore lands

---

[18] "Coal — Bridge to the Future" (1980). It was prepared under the direction of Carroll L. Wilson of the Massachusetts Institute of Technology and is generally referred to as the "World Coal Study." Funding for the study came from private foundations and the Energy Department.

[19] Last year, Congress considered legislation that would have subsidized conversion of power plants to coal. The premise was that utilities could not afford to meet clean air standards for coal emissions. But according to the National Coal Association, half of the plants that would have been eligible for such assistance over a 10-year period already had converted to coal — without federal aid — by last June.

Western Coal — A Special Case

Coal mining in the Western states has become one of the most controversial aspects of the whole issue of the development of a coal economy. The West possesses vast deposits of coal, technically called sub-bituminous. It is of lower heat value than the coal in the East, but it contains less sulfur, which means it burns cleaner than Eastern coal. In addition, it can be easily strip-mined.

Strip mining, already under way at the time of the 1973-74 Arab oil embargo, boomed in its aftermath. Talk of further — and accelerated — coal development has created much disquiet in the West, not only among traditional environmentalists but also among ranchers who see the despoliation of grazing land. A 1977 federal law requires reclamation *(see p. 16)*, but there is doubt that strip-mined land can be effectively reclaimed for grazing in the arid West.

The West's scarcity of water relates to strip mining in another way. Western coal is far from market, making transportation expensive. To solve that problem, electric utilities have developed large generating plants at the mouth of strip mines so they can burn the coal as it emerges from the ground. But these plants require large amounts of water, as do other alternatives such as coal slurry pipelines and coal gasification and liquefaction.

ravaged by uncontrolled mining operations in the past, protected certain lands regarded as unsuitable for surface mining, established mining and mineral resource institutes and provided funds for coal research laboratories and energy graduate fellowships. Many Republicans have said they would like to reduce the powers of the government's Office of Surface Mining and give state governments more regulatory authority over strip-mining operations.[20]

Nuclear energy, in stark contrast to coal, has attracted little or no new investment in recent years. Utilities have placed no new orders for atomic power plants in several years, and last year orders for 10 plants were cancelled. Even so, nuclear power's share in U.S. energy production will increase at a modest pace in the coming years, as plants currently under construction are completed. At the end of 1980, 75 nuclear power plants were licensed to operate in the United States and construction permits had been issued for 85.

President Reagan has made it clear that he wants to encourage greater reliance on nuclear energy. But he has not made any specific statements about how he will do this, and some of the key ideas endorsed by administration officials could face strong opposition in Congress and the states.

---

[20] Another obstacle to much wider use of coal, namely an alleged buildup of carbon dioxide in the atmosphere from increased burning of fossil fuels, so far has played little role in the deliberations of policy makers.

Energy Secretary Edwards said during his confirmation hearings that the federal government should be able to force states to accept nuclear waste disposal sites because governors otherwise would have to take full responsibility for allowing the unpopular sites to be built. This proposal would seem to conflict with the concept of states' rights, which many Republicans hold dear. Edwards also has advocated legislation to streamline the permitting process for nuclear power plants. This, too, could be interpreted as a federal infringement of local rights.

There is one area of nuclear policy where Reagan almost certainly will be able to have a strong impact almost from the start — regulation. During most of the Carter years, the five-member Nuclear Regulatory Commission was split rather evenly between pro-nuclear members and members who took a relatively critical attitude toward the industry's performance. Commissioners Victor Gilinsky and Peter A. Bradford often voted against the position endorsed by the industry, while Joseph M. Hendrie and Richard T. Kennedy usually voted with the industry. The fifth member, John F. Ahearne, was regarded as the swing vote.

Kennedy retired last year, and Carter appointed Albert Carnesale — a Harvard professor — to fill the slot and serve as chairman. But the Senate did not act on this nomination before Carter's term as president expired. Reagan is expected to submit a new name early this year. It is anticipated that he will pick someone to act as a forceful spokesman for the industry's interests.

Joseph Hendrie's position will open up in June, and Reagan will have the opportunity to appoint another member. Despite the two appointments, the effect on the overall balance of power may be less than expected since Reagan will be replacing the two pro-nuclear commissioners with two other pro-nuclear advocates. Even so, most analysts expect the commission to become more pro-nuclear in the coming years.

Prospects for Oil and Gas Exploration

U.S. oil production peaked in 1970, and in the following years the country's petroleum imports rose steeply *(see box, p. 14)*. Many oil analysts believe that rising U.S. demand for imported oil was an important factor in creating a world "seller's market" for oil, which in turn was an important precondition for the success of the Organization of Petroleum Exporting Countries (OPEC) in controlling international price levels.[21] Since the United States has been explored for oil much more intensively than most other parts of the world, geologists

---

[21] See Bruce R. Scott, "OPEC, the American Scapegoat," *Harvard Business Review,* January-February 1981, pp. 6-30.

The Plutonium Economy

One of the most contentious questions facing the Reagan administration and Congress is what to do about certain advanced nuclear technologies, notably the "breeder reactor" and "reprocessing." Reprocessing involves the extraction of reusable uranium and plutonium from used nuclear reactor fuel. Recovering the uranium and plutonium would help stretch nuclear fuel resources, but it also would make plutonium — which can be used as the explosive material in atomic bombs — much more readily available.

Development of a breeder reactor, an advanced reactor that creates new fuel while generating electricity at the same time, would help stretch nuclear fuels even more. But breeder fuels must be reprocessed, and introduction of the breeder would also make plutonium much easier to obtain.

Because of concern that the spread of plutonium technologies would make it easier for terrorists or foreign governments to get the material for atomic bombs, President Carter announced soon after taking office that he would block government funding for a breeder reactor that was to be built at Clinch River, Tenn. Carter was unable to persuade Congress to terminate the Clinch River project, but he did manage to get funding for the project cut to a minimum, and he kept the one U.S. reprocessing plant from opening. The plant is located at Barnwell, S.C., the home state of the new secretary of energy, James B. Edwards *(see box, p. 5)*.

Foreign countries — notably France, Russia and West Germany — have not terminated research and development work on the plutonium technologies, as Carter urged them to do. As a result, pressure has been mounting for the United States to get back into the race to be first to commercialize reprocessing and breeder technologies. Many experts on nuclear technology say, however, that the Clinch River project is a loser and should not be continued even if the United States embarks on a big breeder development program.

and petroleum economists generally are skeptical as to whether there is much more oil to be found in this country.[22]

This view is not shared by some of Reagan's top energy advisers. Michael T. Halbouty, a geologist and independent oil producer who served as the head of Reagan's energy transition team, thinks there is still plenty of oil to be exploited in the United States. "The potential in the United States is not only vast but promising — not only for the next 20 years but well into the 21st century," Halbouty wrote in *The Wall Street Journal*.[23]

Higher prices for petroleum appear to have stimulated more

---

[22] As of 1978, roughly three-quarters of the oil wells drilled in the world were in the United States.
[23] *The Wall Street Journal*, Dec. 27, 1979.

intense exploration, as the advocates of decontrol predicted. Preliminary estimates for 1980 indicate that the number of oil and gas wells drilled was about 23 percent higher than in 1979, and 29 percent higher than in 1978 *(see box, p. 9)*. Domestic oil production in 1980 did not significantly surpass 1979 and 1978 levels, however, and reserves have continued to decline.[24] The amount of natural gas discovered in 1980 may have exceeded consumption for the first time since 1965, and numerous articles have appeared in recent months suggesting that gas is the wave of the future.[25] But according to Douglas Martin, energy reporter for *The New York Times,* "this new attitude toward gas, fostered by people with vested interests, remains distinctly a minority view."[26]

Oil imports, after peaking in 1977, declined modestly in 1978 and 1979 *(see box, p. 14)*. Preliminary estimates for 1980 suggest that they dropped sharply last year to about 6.2 million barrels a day — roughly a 21 percent decline from the preceding year. A decline in domestic consumption, not an increase in production, appears to account for the decrease in dependence on petroleum imports.

Energy Independence Feasibility

B EST guesses about the net impact of decontrol indicate that there will be continuing advances in the area of energy conservation, and that alternatives to oil and gas — coal, solar and nuclear, in roughly that order — will make growing contributions to the U.S. energy supply. But energy analysts also believe that oil and gas will remain crucial for certain activities in the foreseeable future, and that the United States will remain substantially dependent on foreign supplies of petroleum.

Because of the country's continuing dependence on foreign oil and its vulnerability to embargoes and shortages, the Carter administration came under a lot of criticism last year for not acting more expeditiously to fill the strategic petroleum reserve — oil stored underground in salt domes in Louisiana and Texas. Legislation passed in 1975 authorized establishment of a reserve containing 750 million to 1 billion barrels of oil. But as of late 1980, the reserve contained just a little over 90 million

---

[24] In the first 10 months of 1980, domestic oil production averaged 8.6 million barrels a day, compared to 8.5 million in 1979 and 8.7 million in 1978. See Energy Department's *Monthly Energy Review,* December 1980, p. 26.
[25] See for example Stephen Chapman, "The Perfect Fuel," *The New Republic,* Nov. 22, 1980, pp. 11-14, and Gregg Easterbrook, "Natural Gas," *The Washington Monthly,* October 1980, pp. 20-32.
[26] Writing in *The New York Times,* Dec. 1, 1980.

barrels, and for roughly a year or so no oil had been added.

The Federation of American Scientists, among others, charged that the Carter administration's "reluctance" to fill the reserve was "an exaggerated response to not very strenuous Saudi [Arabian] objections to our filling the reserve."[27] An amendment to the Energy Security Act of 1980 ordered the executive branch to fill the reserve at a rate of at least 100,000 barrels a day, and about 130,000 barrels are currently being added each day. The reserve now contains 110 million barrels. The Reagan administration has promised to move quickly on filling the reserve.

There is likely to be some discussion as the reserve is filled about the conditions under which oil would be released for consumption. According to one school of thought, which probably will prevail among Reagan officials, the oil should be released only in a national emergency — i.e., in the face of a foreign oil embargo or, at most, an exceptionally acute shortage. It has been suggested, however, that the reserve could be used on an ongoing basis to stabilize the cost of energy in the home market.

Dependence of Allies on Persian Gulf Oil

With the repeated interruptions and curtailments of oil exports from the Persian Gulf during 1979 and 1980, it became evident to many Americans that making the United States relatively invulnerable to oil cutoffs would not completely solve the problem of "energy security." During the Nixon, Ford and Carter administrations, officials generally said that achievement of "energy independence" was the country's overriding goal — to minimize the risk of "oil blackmail" by Middle Eastern suppliers.

After the Soviet Union invaded Afghanistan, fears spread that Russia might move to cut off oil supplies to Europe and Japan. Western Europe gets roughly 60 percent of its oil from the Persian Gulf, and Japan roughly 70 percent. In response to these fears, Carter said in his State of the Union address on Jan. 23, 1980, "An attempt by any outside force to gain control of the Persian Gulf region will be regarded as an assault on the vital interests of the United States of America and . . . will be repelled by any means necessary, including military force."

This "Carter Doctrine" raised a number of contentious issues that have yet to be fully resolved.[28] Some critics have said

---

[27] *F.A.S. Public Interest Report,* November 1980, p. 2.
[28] See "Foreign Policy Issues in the 1980 Campaign," *E.R.R.,* 1980 Vol. I, pp. 205-224.

that the United States does not have the military power to enforce the doctrine in an area so close to Russia and so far from the United States. During his Senate confirmation hearings, Reagan's secretary of defense, Caspar W. Weinberger, described the Carter Doctrine as "extraordinarily clumsy and ill-advised." Other critics have wondered whether the United States should commit itself to defending Europe's and Japan's energy supplies by force of arms, especially in view of their lukewarm support for U.S. policies in the Middle East. Still others have asked why the United States has been paying such a high price to achieve "energy independence," if in the end Americans might have to fight wars to defend oil supplies anyway.

Proposals for a Global Energy Strategy

A number of prestigious analysts recently have stressed the dangers of making reduced imports the be-all and end-all of U.S. energy policy. A Harvard research project, involving several specialized branches of the university, has issued a report that draws attention not only to U.S. economic and military inter-dependence with Europe and Japan but also to "[our] vital economic, foreign-policy, security and humanitarian concerns in developing nations around the world."[29] Soaring oil import bills have caused the foreign debts of many developing countries to rise sharply in the 1970s, and their growing debt service burden now threatens economic growth and social stability. Among the countries most seriously affected by such problems are some that have close economic and political relations with the United States — Brazil, South Korea and the Philippines, for example.[30]

The Senate Committee on Energy and Natural Resources on Nov. 20, 1980, issued a report in which the staff concluded that "[a]n energy policy aimed solely at reducing imports will not adequately insulate the United States from the gathering energy crisis. . . . [W]e must recognize that the energy problem has many international aspects. As a consequence, our foreign and defense policies must be considered as much a part of our overall energy policy as our traditional programs to reduce oil imports."[31]

Writing in *Foreign Affairs* magazine last summer, Walter J. Levy, a leading oil consultant, said that ways must be found of keeping the real price of oil at a "manageable level." Levy

[29] For a summary, see David A. Deese and Joseph S. Nye, "Energy and Security," *Harvard Magazine,* January-February 1981, pp. 40B-40H.

[30] See "Third World Debts," *E.R.R.,* 1980 Vol. II, pp. 541-560, and "The Philippines Under Stress," *E.R.R.,* 1980 Vol. II, pp. 757-776.

[31] Executive Summary reprinted in *Science* magazine, "The Geopolitics of Oil," Dec. 19, 1980, pp. 1324-1327.

said the "underlying concept would have to be that oil must be conceived in terms of a 'common heritage of mankind.' " Conceding that this might be a "utopian fantasy," Levy said controlling the oil price would require "an accommodation between oil producers and oil importers where both parties take full account of each other's vital interests. Above all, it presupposes that the United States, the Soviet Union and possibly China agree on common principles that would apply to their policies affecting the area [the Persian Gulf], and thus would be willing to abstain from any action that might interfere with the maintenance or establishment of peace and tranquility in the Middle East."[32]

It remains to be seen what the Reagan administration will do to stabilize the Persian Gulf, reassure allies about their supplies and help prevent Third World countries from running up against insoluble economic problems. Whatever Reagan does, he will be getting some of his most important advice on energy policies not only from the Energy Department, but from the State Department and the Pentagon as well.

---

[32] Walter J. Levy, "Oil and the Decline of the West," *Foreign Affairs,* summer 1980, p. 1014.

Selected Bibliography

Books

Congressional Quarterly, *Energy Policy,* 2nd edition (forthcoming), March 1981.

Daly, Charles, ed., *Energy Security: Can We Cope with the Crisis?,* American Enterprise Institute, 1981.

Davis, David Howard, *Energy Politics,* St. Martin's Press, 1978.

Goodwin, Craufurd, *Energy Policy in Perspective: Today's Problems and Yesterday's Solutions,* The Brookings Institution, 1981.

Stobaugh, Robert and Daniel Yergin, *Energy Future,* Random House, 1979.

Articles

Bethel, Thomas N., "Synfuels," *The Washington Monthly,* October 1980.

Energy Daily (Washington), selected issues.

Howard, Niles, "The New Glow in Solar Electricity," *Dun's Review,* November 1980.

Menard, H. William, "Toward a Rational Strategy for Oil Exploration," *Scientific American,* January 1981.

Petroleum Intelligence Weekly (New York), selected issues.

Reilly, Ann M., "The Big Battles over Energy," *Dun's Review,* January 1981.

Reports and Studies

Department of Energy, "Reducing U.S. Oil Vulnerability," DOE/PE/0021, Nov. 10, 1980.

Editorial Research Reports: "Nuclear Fusion Development," 1980 Vol. II, p. 657; "Public Confidence and Energy," 1979 Vol. I, p. 38; "Auto Research and Regulation," 1979 Vol. I, p. 145; "Oil Imports," 1978 Vol. II, p. 621; "America's Coal Economy," 1978 Vol. I, p. 281; and "Synthetic Fuels," 1979 Vol. II, p. 621.

General Accounting Office, "Oil and Natural Gas from Alaska, Canada and Mexico — Only Limited Help for U.S.," Washington, Sept. 11, 1980.

National Research Council, Committee on Nuclear and Alternative Energy Systems, "Energy in Transition," January 1980.

Senate Committee on Energy and Natural Resources, "The Geopolitics of Oil," November 1980. Executive summary reprinted in *Science* magazine, Dec. 19, 1980.

World Coal Study, "Coal — Bridge to the Future," published by Ballinger, 1980

Cover art by Staff Artist Robert Redding

AMERICA'S NUCLEAR WASTE BACKLOG

by

Robert Benenson

**Dec. 4
1981**

Editor's Note: Both houses of Congress are still grabbling with nuclear waste legislation and there is no certainty that a comprehensive waste disposal bill will be passed this year. The primary areas of disagreement continue to be whether the government should provide spent-fuel storage for nuclear utilities *(see p. 30)*, how to deal with military waste *(see p. 39)*, and what will be the role of the states in determining the location of waste repositories *(see p. 38)*.

A federal district court judge in New York City on Feb. 20, 1982, blocked a new Department of Transportation rule that would have allowed nuclear waste to be trucked through major urban areas. Judge Abraham Sofaer, ruling on a suit brought by the city and state of New York, said "the failure of DOT adequately to assess the likelihood and consequences of accidents, human error and malevolence made the agency incapable of justifying a routing rule for spent fuel and large-quantity radioactive materials that overrides non-federal regulation aimed at avoiding those risks in densely populated areas." New York and other cities have local bans against transporting nuclear waste.

AMERICA'S NUCLEAR WASTE BACKLOG

O N DEC. 2, 1942, a team of scientists led by physicist Enrico Fermi produced the first controlled atomic reaction and launched the "nuclear age." The products of nuclear technology, particularly nuclear weapons and nuclear power generation, have changed the world. They have also bequeathed a dubious legacy to future generations — 40 years of accumulated radioactive waste. The problem of nuclear waste disposal does not grab the headlines as often as nuclear weapons proliferation or nuclear power plant safety issues.[1] Nonetheless, nuclear waste is one of the most perplexing environmental problems of our time. After almost four decades, the United States still lacks a national plan for nuclear waste disposal.

The first decades of the nuclear age were marked by a distinct lack of urgency concerning the disposal of nuclear waste. The emphasis was on production of nuclear weapons and nuclear power. Assured by scientists that a safe, technical solution to the nuclear waste disposal problem was at hand, government authorities permitted wastes to collect in temporary storage facilities. Only after the temporary facilities became pressed for space, and the public's concern for the environment spread to nuclear waste, did the government accelerate its search for a permanent answer to the nuclear waste question.[2]

When President Carter unveiled his plan for managing high-level radioactive wastes in February 1980, it was the first time that a U.S. president had advanced a comprehensive strategy for nuclear waste disposal.[3] But the Carter administration did not succeed in winning congressional approval of the president's program.

When President Reagan publicly addressed the nuclear waste issue for the first time on Oct. 8, 1981, his statement was

---

[1] See "Controlling Nuclear Proliferation," *E.R.R.*, 1981 Vol. II, pp. 509-532 and "Nuclear Safety," *E.R.R.*, 1975 Vol. II, pp. 601-624.
[2] See "Nuclear Waste Disposal," *E.R.R.*, 1976 Vol. II, pp. 883-906.
[3] In 1978 Carter established an Interagency Review Group to examine the nuclear waste problem. It made its recommendations in March 1979. However, the Carter administration did not send its legislative proposal to Congress until Feb. 12, 1980. Carter's plan called for permanent burial of spent nuclear fuel in an underground rock deposit by the mid-1990s; establishment of a state planning council to deal with nuclear waste problems; federally funded away-from-reactor storage facilities for spent fuel that utilities did not have room for; and extension of the Nuclear Regulatory Commission's licensing authority to cover low-level waste storage.

more of a policy directive than a plan for action. Reagan said that he had instructed Secretary of Energy James B. Edwards "to proceed swiftly toward deployment of means of storing and disposing of commercial high-level radioactive waste." But Reagan also reconfirmed his support for an expanded nuclear power program,[4] a policy that could exacerbate the nuclear waste problem, and he did not mention the status of military nuclear waste, which amounts to over 90 percent, by volume, of the nation's high-level radioactive waste inventory *(see p. 39)*.

Congress is now struggling to devise a program for the disposal of the most dangerous nuclear waste products.[5] Both houses of Congress are considering various pieces of legislation, and final action is not expected before summer 1982. The previous Congress came close, but failed to pass a high-level nuclear waste disposal bill in December 1980 *(see p. 41)*. Crafting nuclear waste legislation is difficult because it requires the careful balancing of the conflicting needs and desires of so many competing interest groups: the nuclear power industry, the military, environmentalists, and local officials and residents who may support the principle of nuclear waste disposal as long as the waste is placed in someone else's backyard.

Urgency of the Waste Disposal System

The snail's pace at which a national nuclear waste policy is being developed belies the urgency of the radioactive waste disposal problem. High-level liquid waste, containing such elements as strontium and cesium, remains intensely radioactive for between 600 and 1,000 years. Transuranic wastes *(see box, p. 29)* such as plutonium and americium maintain their toxicity for up to 500,000 years. Spent fuel assemblies from commercial power plants, the deadliest of all radioactive waste products, contain both high-level and transuranic elements.

All of these extremely dangerous wastes, currently housed in temporary storage facilities, must eventually be isolated from the human environment for hundreds or thousands of years. Decisions made today on nuclear waste disposal could leave an indelible mark on the future. "Once entered on their winding course through the environment, radioactive isotopes are out of reach of man's control," Sheldon Novick wrote in *The Careless Atom* (1969). "The damage, once done, is irremediable."

Nuclear advocates dismiss these concerns as apocalyptic, citing scientific studies that favorably compare the safety of radio-

---

[4] Reagan called nuclear power "one of the best potential sources of new electrical energy supplies in the coming decades."

[5] Congress has dealt with other forms of nuclear waste, passing legislation concerning uranium mill tailings in 1978 *(see box, p. 29)* and low-level waste dumping in 1980 *(see box, p. 40)*.

Types of Radioactive Waste

Mill tailings are fine sands resulting from the refinement of uranium ore and containing naturally occurring radioactive materials. The most voluminous of all radioactive wastes — over 3 billion cubic feet have been dumped at 21 inactive and 24 active mill sites — uranium tailings emit radon gas, which many scientists believe to be carcinogenic. Mismanagement of these wastes resulted in congressional passage of the Uranium Mill Tailings Control Act of 1978, which provided for a federal-state cleanup of abandoned tailings sites. But Congress still has not appropriated funds for the program.

Low-level wastes are generated by all activities using radioactive materials. They range from contaminated protective clothing and tools to industrial and medical wastes. The term "low level" refers not to the degree of radioactivity but to the source. Basically, all wastes not produced in nuclear reactors or in the reprocessing of nuclear fuel are classified as low-level wastes. Congress passed a law in 1980 making low-level waste management a state responsibility *(see box, p. 40)*.

Transuranic wastes, such as plutonium, neptunium and americium, are produced in the reprocessing of nuclear fuel and the production of nuclear bombs. These elements can be dangerous if inhaled or ingested, and they remain toxic for hundred thousands of years, far longer than high-level wastes. Most of the 2 million cubic feet of transuranic wastes that are in retrievable storage are housed at the Idaho National Engineering Laboratory at Idaho Falls.

Spent fuel assemblies. Creation of fission products, such as strontium and cesium, and byproducts, such as plutonium, blocks the nuclear reaction and forces periodic replacement of the fuel assemblies. Thousands of these spent fuel assemblies are temporarily stored in water cooling "pools," awaiting final decision as to whether to reprocess them or dispose of them as waste.

High-level waste. When spent fuel is reprocessed to remove remaining fuel elements, a highly-radiocactive liquid waste is created. These wastes, containing dangerous, penetrating elements such as strontium and cesium, must be isolated from the environment for hundreds of years. Most of the nation's 77-million-gallon inventory of high-level waste was created in defense programs, and is stored at Hanford, Wash., Aiken, S.C., and Idaho Falls, Idaho. The only commercial high-level wastes, some 600,000 gallons, are stored at West Valley, N.Y.

Decontamination-and-Decommissioning Wastes. The waste problem of the future. Nuclear reactors and other nuclear facilities are built with a lifespan of only about 40 years. After they outlive their usefulness, these facilities must be dismantled and treated as radioactive waste. Some nuclear facilities are already nearing the point of obsolescence.

active waste with other forms of industrial waste. "In any scientifically based listing of threats to human health in our society, radioactive waste would be very far down on the list," University of Pittsburgh physics Professor Bernard L. Cohen told *Nuclear Industry* magazine.[6] "It is thousands of times less damaging to human health than the disposal of wastes from burning fossil fuels...." The March 1979 issue of *Fortune* magazine reported: "The industrial world routinely uses hundreds of other dangerous substances whose toxicity, far larger in aggregate, stays at full strength forever — for example, mercury, arsenic, and some poisonous compounds."[7]

Such optimistic statements cannot erase the fact that increasing nuclear waste inventories are causing logistical, economic and public relations problems for the nuclear industry, which is actively advocating the passage of waste legislation. The urgency of the situation was underscored by Sen. Alan Simpson, R-Wyo., chairman of the Environment and Public Works Subcommittee on Nuclear Regulation. "What will bring nuclear power to its knees," Simpson said, "is the failure to deal with the waste products of that industry...."[8]

Spent Fuel and Debate Over Reprocessing

The most pressing waste disposal problem concerns spent fuel assemblies from commercial nuclear power plants. The fission reaction that helps create electricity also creates fission products, such as strontium and cesium, and byproducts, such as plutonium. Eventually, these waste products block the fission process, rendering the fuel assemblies unusable. The obsolete assemblies must be replaced, but when removed from the reactor, they are far more radioactive than when they were put in.

In order to contain the radioactivity in the spent fuel assemblies, each commercial reactor was built with an adjoining water pool into which the assemblies are placed to "cool." These storage pools were constructed with limited capacities because they were intended to be temporary. From the inception of the commercial nuclear energy program in 1954, it was expected that commercial spent fuel would eventually be removed to commercial "reprocessing" plants.

Reprocessing is a chemical process which strips valuable fuel elements from the spent fuel assemblies, leaving behind a high-level liquid waste. The defense nuclear industry recovers plutonium through reprocessing for use in nuclear weapons. Re-

---

[6] *Nuclear Industry*, January 1981.
[7] Edmund Faltermeyer, "Burying Nuclear Trash Where It Will Stay Put," *Fortune*, March 26, 1979, p. 98.
[8] Simpson made these remarks during an appearance on "The MacNeil-Lehrer Report," PBS-TV, July 5, 1981.

Storage Sites for Commercial Nuclear Waste (Existing and Potential)

- 138 commercial nuclear reactors already operating or expected to be in operation by 1985. Spent fuel is stored at sites.

★ Potential away-from-reactor regional storage sites, under consideration by Department of Energy (Morris, Ill.; West Valley, N.Y.; Barnwell, S.C.)

⊙ Existing commercial low-level waste sites (Beatty, Nev.; Hanford, Wash.; Barnwell, S.C.). In addition, there are 7 major and 7 minor low-level waste sites used by federal government.

☐ 23 states being considered for high-level nuclear waste storage in geological formations.

SOURCE: U.S. Department of Energy

processed plutonium and uranium could also be utilized as fuel for commercial reactors. At one time, government and industry officials were convinced that commercial fuel reprocessing would be both necessary and profitable because of an anticipated shortfall in uranium supplies and the expected development of plutonium-consuming "breeder reactors."

But demand for nuclear energy has fallen way short of projections, leaving uranium fuel in relatively ample supply. And breeder reactor technology remains grounded as controversy rages over its high cost and the necessity of a reactor that creates more plutonium than it uses. Lacking the demand for its services and crippled by escalating costs, the commercial reprocessing business has been a bust. Only one commercial processing plant, at West Valley, N.Y., ever opened, and it closed in 1972 after only six years of operation. Another reprocessing project, at Morris, Ill., was never completed because of design problems.

A third plant was nearing completion at Barnwell, S.C., in 1977, when President Carter banned commercial reprocessing. Like his predecessor, Gerald R. Ford, Carter expressed concern that reprocessing of commercial fuel would set an example for other nations, which might divert recovered plutonium for use in nuclear weapons. Carter also worried that growing supplies of weapons-grade plutonium would provide an enticing target for terrorist groups.[9]

President Reagan seems intent on reviving the reprocessing option. In his Oct. 8 statement on nuclear power, Reagan announced: "I am lifting the indefinite ban which previous administrations placed on commercial reprocessing activities in the United States." Reagan also promised to "eliminate regulatory impediments to commercial interest in this technology."

Reprocessing of spent commercial fuel has its advantages. If the United States proceeds with an extensive nuclear power program, and especially if the breeder reactor becomes a reality, the fuel elements of spent commercial assemblies will become extremely valuable. Without reprocessing, these fuel elements would have to be disposed of as waste. As for fears about nuclear proliferation, reprocessing advocates say that a viable commercial reprocessing industry in the United States could contract for the spent fuel of other nations, thereby giving the U.S. more control over the future disposition of the plutonium and other reprocessing products.[10]

The feasibility of commercial reprocessing in the United States is contingent on the expansion of the nuclear power industry. Today, with expenses high and prospects for profit low, there are no known entrepreneurs planning to plunge into the commercial reprocessing business. Although President Reagan supports reprocessing, it is unlikely that his budget-cutting administration will come up with federal dollars to prop up the dormant commercial industry.

Certain officials in the Reagan administration and in the defense and energy establishments believe that a domestic commercial reprocessing industry could be revived if the government would change its longstanding policy prohibiting use of commercial nuclear waste for defense weapons purposes. "What the hell difference does it make where the plutonium comes from?" asked House Armed Services Committee staffer Seymour Schiller. "The country is going to have a nuclear weapons program. It's just a question of doing it economically."

This position brought a sharp response from Sen. Gary Hart, D-Colo. Warning against diversion of civilian waste to military use, Hart said, "No longer would the United States have the credibility necessary to persuade currently non-nuclear-weapons countries not to use their ostensibly peaceful nuclear power

---

[9] See "International Terrorism," *E.R.R.*, 1977 Vol. II, pp. 929-931, and "Nuclear Proliferation," *E.R.R.*, 1978 Vol. I, pp. 201-220.

[10] France already has a commercial reprocessing industry, and several other nations, including Great Britain, Belgium and Spain, are still considering the reprocessing option.

programs to build atomic bombs."[11] When the Senate Environment and Public Works Committee reported a nuclear waste disposal bill on Nov. 16 *(see p. 42)*, it included an amendment introduced by Hart that would prohibit the use of spent commercial fuel for military purposes.

Options For Storing Spent Atomic Fuel

While the battle over reprocessing versus disposal of commercial spent fuel rages, the spent fuel assemblies continue to pile up at the nation's 78 commercial nuclear power plants. At the end of 1980, there were 28,315 stored assemblies, with another 5,775 expected to be added this year. Some of the storage pools are nearing their capacities; several may run out of space as early as 1986. The problem may be deferred temporarily by re-racking (moving the fuel assemblies closer together in the storage pools) or transshipment (shipping spent fuel assemblies to plants with spare storage capacity). Even reshuffling has its limits, though, and alternative storage or disposal facilities will eventually be required.

There are several possible solutions to the spent fuel storage problem. "Dry storage" technology, a space-efficient alternative to cooling pools, is currently under development, but is not expected to be perfected and available for several years. Power plant owners could add on to their present storage capacities, but this is an expensive option that financially strapped utilities would prefer to avoid.

An option that has a good deal of industry support is a federally funded, away-from-reactor (AFR) storage facility. The AFR concept was first broached by the Atomic Energy Commission in the early 1970s. AEC's successor agency, the Energy Research and Development Administration, shelved the idea in 1975. President Carter revived the AFR concept in 1980. President Reagan has stated his opposition to federal financing of a commercial storage facility, but a bill currently under consideration in the Senate contains AFR authorization.

The away-from-reactor facility would be intended for short-term storage, until a solution to the disposal problem can be implemented. Another possibility is a long-term, monitored retrievable surface storage (MRSS) facility. The advantage of such a long-term storage facility is that it would provide more time for decision-making concerning reprocessing and permanent disposal site selection. Both the surface storage plan and the away-from-reactor storage idea meet with vocal opposition from environmentalists, who see interim storage as a stopgap measure that will delay final disposition of the spent fuel, and

---

[11] Hart and Schiller were quoted in *Congressional Quarterly Weekly Report*, Nov. 7, 1981, p. 2183.

from officials and local residents of areas mentioned as potential AFR/MRSS sites.

High-Level Waste Storage

I F THE reprocessing option is never exercised, commercial spent fuel assemblies will eventually have to be isolated and disposed of permanently. But even if commercial spent fuel is reprocessed, it will just add to the nation's huge inventory of high-level liquid radioactive waste. High-level liquid waste is laced with deadly amounts of strontium, cesium and other radioactive elements. These elements emit highly penetrating alpha and beta radiation, which can cause somatic and genetic damage to humans and animals. They remain toxic for centuries, and must be quarantined from the human environment.

These liquid wastes, like the spent fuel assemblies, are currently stored in temporary facilities. Most of these wastes originated in the reprocessing of spent fuel from the defense weapons program. These defense high-level wastes are housed in tanks at federal nuclear reservations run by the Department of Energy (DOE).

The current high-level liquid waste inventory is about 77 million gallons.[12] Of this total, 48 million gallons are stored in tanks at the Hanford nuclear reservation at Richland, Wash. Another 26 million gallons are stored at the Savannah River plant near Aiken, S.C. Three million gallons are stored at the Idaho National Engineering Laboratory near Idaho Falls. Most of the current defense reprocessing activity is taking place at the Savannah River plant, where 1.3 million gallons of high-level waste are added each year.

The only commercial high-level wastes ever produced are stored at the defunct reprocessing facility at West Valley, N.Y. The West Valley plant, in operation from 1966 to 1972, produced 600,000 gallons of high-level waste. In 1980, Congress brought the West Valley site under federal supervision, certifying an agreement by which the federal government will pay 90 percent of the cost of cleaning up the West Valley wastes, with New York state paying the remaining 10 percent. The Department of Energy plans to use the West Valley facility as a pilot project to test and develop the technological procedures for solidification and transportation of the dangerous liquid wastes (see p. 35).

---

[12] A small percentage of this waste has been solidified into sludge or calcine powder.

Along with the high-level wastes, disposal must be arranged for transuranic wastes, or wastes that contain small amounts of elements that are heavier than uranium, such as plutonium and neptunium. Although these wastes emit less-penetrating radiation and must be inhaled or ingested to cause serious injury, they have much longer toxic lives than high-level wastes — in the hundreds of thousands of years. Around 2 million cubic feet of these transuranic wastes, which result from the production of nuclear weapons and the reprocessing of nuclear fuel, are currently in retrievable storage, most of them at the Idaho National Engineering Laboratory.

Storage Options For Long-Lived Wastes

High-level liquid radioactive wastes cannot be transported in their current, volatile form. The predominant scientific strategy for disposal is to solidify the wastes, either by mixing them with glass or turning them to calcine powder, encapsulating them in glass or concrete containers, and burying them in deep, stable geological formations.

Nuclear industry officials, relying on a large body of scientific research, are confident that a secure final resting place can be found for the high-level wastes. In its report to President Carter in March 1979, the Interagency Review Group on Nuclear Waste Management[13] said: "No scientific or technical reason is known that would prevent identifying a site that is suitable for a repository, provided that the systems view is utilized rigorously to evaluate the suitability of sites and designs, and in minimizing the influence of future human activity."

A high-level waste repository must be capable of isolating wastes that will remain dangerous for thousands of years. The lifetimes of these wastes far exceed the average longevity of social institutions as we know them, so waste repositories must be secure enough to protect the environment even in the absence of traditional maintenance procedures. High-level wastes must be disposed of hundreds of feet underground, in geological formations that are not vulnerable to earthquakes (to prevent escape of radiation through faults), do not contain valuable resources (so future generations will not accidentally dig up the wastes), and are waterproof (to keep radioisotopes from leaching into potable groundwater).

Salt has long been considered the most suitable medium for high-level and transuranic waste disposal. Many scientists believe that the presence of deep, solid salt formations indicates the absence of faults and the long-term absence of water, since the salt would otherwise have long since dissolved. As early as

---

[13] President Carter established the Interagency Review Group in March 1978 to examine the nuclear waste disposal problem *(see footnote 3)*.

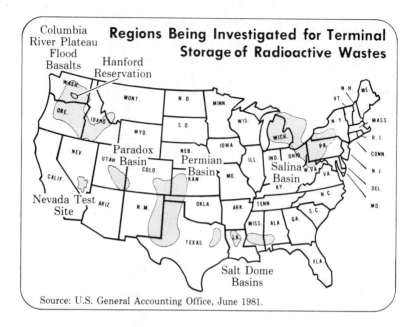

Regions Being Investigated for Terminal Storage of Radioactive Wastes

Columbia River Plateau Flood Basalts

Hanford Reservation

Paradox Basin

Permian Basin

Salina Basin

Nevada Test Site

Salt Dome Basins

Source: U.S. General Accounting Office, June 1981.

the mid-1950s, scientists thought they had located a prime spot for nuclear waste disposal: an abandoned salt mine near Lyons, Kan. Despite local political opposition, exploratory digging and research was begun at the Lyons site in 1971, but it was discontinued shortly thereafter. Thousands of gallons of water that had been pumped into the ground by a mineral development company working near the research site mysteriously disappeared, indicating the existence of an underground fissure and casting doubts upon the safety of the venture.

Search for Suitable Geologic Formations

Since the Lyons failure, the search has continued for a suitable storage site. The Department of Energy wants to accelerate a long-delayed plan to open a Waste Isolation Pilot Project (WIPP) in salt beds near Carlsbad, N.M., for storage of transuranic wastes and for research and development activities on high-level waste storage. Salt beds and salt domes in Texas, Utah, Mississippi, Louisiana, Kansas and Nevada are also being eyed as potential high-level waste repositories. Since the Lyons, Kan., incident raised questions about the acceptability of salt as a nuclear waste disposal medium, government researchers are investigating other geological formations as well. These include basalt in the Pacific Northwest, welded tuffs (volcanic ash formations) in the Nevada desert, and granite in the Lake Superior region and in the Appalachians.

The timetable for opening a deep geological repository is one of the issues that will have to be dealt with by Congress. Generally, suggested timetables follow a pattern. The Department of Energy will conduct research activities at proposed repository sites, leading to the recommendation of one or more sites, to be followed by testing and by consultation with affected states. After several years of testing, the DOE would turn the results over to the president, who would recommend a repository site to Congress. After congressional approval, development of a full-scale repository would begin. Nuclear industry officials are confident that a high-level waste repository can be in operation by the early 1990s.

Despite the official optimism, the days of blind faith in science have long since passed, and there are many people who have serious doubts about the proposed solution to the nuclear waste disposal problem. These doubts have made the issue of high-level waste disposal one of American politics' hottest potatoes.

Politics of Waste Disposal

INDUSTRIALISTS often flinch at the mention of waste disposal or environmental problems, but nuclear power advocates are openly lobbying for passage of a nuclear waste disposal bill. Legislation would not only unburden the utilities of their unwieldy garbage disposal problem, but would also solve one of the major public relations dilemmas faced by the industry. "There are many segments of the public that are honestly concerned about the waste," Edwin Wiggin, vice president of the Atomic Industrial Forum, said in a recent interview. "There are many segments of the public that honestly believe that we don't know what to do with it and that we don't have the capability to handle it."

Industry officials believe they are fully capable of handling the waste problem, and they want to spread the word. "That there are no technical barriers to the safe disposal of nuclear waste is a message that both the industry and federal government must communicate to the American public," Sherwood H. Smith Jr., chief executive officer of the Carolina Power and Light Co., said in testimony before a Senate panel.[14]

It will take more than optimistic statements and scientific theories to convince skeptics that there is a safe solution to the growing nuclear waste problem. Foremost among the critics are environmentalists and anti-nuclear activists. Although no in-

---

[14] Smith appeared Oct. 6, 1981, before a joint hearing of the Senate Energy Committee and the Environment and Public Works Subcommittee on Nuclear Regulation.

cident involving nuclear waste in the United States is known to have taken human life, anti-nuclear groups claim that the nation's record for nuclear waste maintenance gives little reason for future optimism. Environmentalists point to the errors of the nuclear past: not only leaking high-level waste tanks at Hanford, but sodden low-level waste trenches leaching radioisotopes into groundwater; barrels of low-level waste, dumped in the oceans, now corroded and leaking; and mountains of radioactive mill tailings, left unsupervised and vulnerable to wind and rain.

Some anti-nuclear activists cast doubt upon the ability of scientists to dispose of long-lived toxic wastes safely. "Even the most brilliant scientist may honestly think he's developed a container that will remain intact for a half million years.... [But] he's going to be dead in 30 years; he will never live to verify his hypothesis," Dr. Helen Caldicott said at a Department of Energy public hearing on nuclear waste management held in Boston in 1978, adding, "You know damn well that we will never know if we are going to store this waste safely."[15]

Other nuclear foes acknowledge the need for permanent disposal of high-level nuclear wastes, but they call for long-term studies to prove that the medium for disposal will safely isolate the wastes for hundreds or thousands of years. Proposed development schedules that foresee a high-level waste repository in operation by the early 1990s are unrealistic, critics say. Sierra Club Washington representative Brooks Yeager said, "It's absolutely clear they're moving too fast."[16]

Since nuclear power and nuclear weapons operations add to the nuclear waste inventory, some anti-nuclear activists have used the waste problem as a pretext to call for halting all nuclear activities. "Among many anti-nuclear activists back at the grass-roots level, their attitude is, 'Yes, we've got this waste problem, waste exists, and the first thing we do to solve the problem is we stop creating the waste,'" Caroline LeGette, legislative representatives for Ralph Nader's Congress Watch organization, said in a recent interview.

Position of State Officials on the Issue

In their efforts to forestall "rushed" legislation on nuclear waste disposal, nuclear opponents have found some unexpected allies — the governors and legislators of many of the 50 states. Except for the few who openly oppose nuclear power, such as Edmund G. "Jerry" Brown Jr. of California, the governors

---

[15] Caldicott was then president of Physicians for Social Responsibility.
[16] Quoted in *The Congress Watcher*, October-November 1981, p. 7. *The Congress Watcher* is published by the Ralph Nader organization Congress Watch.

acknowledge the need for expanded nuclear programs. But when it comes to the disposal of nuclear waste, their attitude is almost universal: "Don't put it here!"

As part of its deliberations on pending nuclear waste bill, the Senate Environment and Public Works Subcommittee on Nuclear Regulation invited comment from governors and officials of states that include potential waste maintenance or disposal sites. The following remarks are typical of the positions taken by those who testified before the committee on Nov. 9, 1981:

> "For over thirty years, the citizens of South Carolina have done more than their fair share to shoulder the nation's nuclear waste burden. . . . South Carolina will not be the path of least resistance in seeking a national solution to a national problem." — Gov. Richard Riley, D-S.C.[17]

> "We are willing to continue to assume a fair and reasonable responsibility for meeting away-from-reactor spent fuel storage needs. However, we will strongly resist assuming the total burden for away-from-reactor fuel storage." — Philip F. Gustafson, Illinois Director of Nuclear Safety.

> "The political sector as well as the Mississippi public have voiced their vehement opposition to the development of a repository within one of the salt domes in one of our southeastern counties." — John W. Green Jr., Nuclear Waste Program manager, Mississippi Energy and Transportation Board.

More than 20 states have passed laws or referenda establishing prohibitions or limits on nuclear waste dumping.[18] In addition, California and Oregon have laws that ban construction of new nuclear power facilities until an acceptable solution is found to the nuclear waste problem. Although a comprehensive national law would probably take precedence over these state bans, most state officials want provision for a state "veto" of a proposed nuclear waste site, which could be overturned only by a majority vote in one or both houses of Congress.

Old Dispute Over Defense-Related Wastes

The 97th Congress appears ready to act on high-level nuclear waste legislation. Even opponents of nuclear expansion admit the necessity of legislation. Rep. Morris K. Udall, D-Ariz., chair-

---

[17] Riley was the chairman of the State Planning Council on Nuclear Waste Management, which was appointed by President Carter in February 1980. The panel expired last summer. Riley's state, South Carolina, already contains the Savannah River high-level waste storage and reprocessing facility, the nation's largest low-level commercial waste facility at Barnwell and the nearly completed Barnwell commercial reprocessing plant.

[18] The following states have laws banning the disposal of all radioactive wastes: Alabama, Maryland, Michigan, Oregon, Indiana and Louisiana. Eight states have banned disposal of high-level wastes only: Connecticut, Illinois, Montana, New Hampshire, South Dakota, Texas, Vermont and West Virginia. The following states require legislative approval for radioactive waste dumping: Colorado, Connecticut, Kentucky, Louisiana, Maine, Minnesota, Mississippi, New Hampshire and North Dakota. The state of Washington had a ban on importation of low-level nuclear waste, but this law, as well as Illinois' ban on importation of spent fuel assemblies, were ruled unconstitutional in federal district courts.

Congressional Action on Low-Level Nuclear Wastes

The management of low-level radioactive wastes has been erratic over the years. Until the early 1970s, low-level nuclear waste usually was dumped in barrels into the oceans. Follow-up studies found that many of the barrels corroded and leaked. There were also problems with low-level wastes buried on land. Many of the disposal sites became contaminated with highly radioactive elements such as plutonium.

Low-level wastes produced by the defense industry are disposed of at 14 federal sites. But only three states — Washington, South Carolina and Nevada — operate commercial low-level waste sites and the residents and governors of those states were threatening to close them down.

Washington voters approved a ballot initiative Nov. 4, 1980, to close the disposal site at Hanford to out-of-state nuclear waste beginning in the summer of 1981. Nevada Gov. Robert F. List said he was considering closing that state's dump at Beatty, and South Carolina Gov. Richard Riley, whose state accepted about 75 percent of the nation's nuclear trash, said it would accept less in the future.

To deal with this situation, Congress in December 1980 passed legislation making disposal of commercial low-level waste a state responsibility. States could either build their own dump sites or form regional compacts with other states to establish a mutual burial site. After 1985, these regional groups, which had to be approved by Congress, can exclude wastes from states that did not join the compact.

man of the House Interior Committee, is a sponsor of one of the nuclear waste bills currently being considered (HR 3809). "We've been accumulating waste now since 1944 and we have yet to dispose of the first pound of the stuff," Udall said in an interview. "The time has come to get going on finding and isolating and developing and building a deep geological storage depository."

The development of needed legislation is snagged, though, because many congressmen are working hard to protect their constituencies ... and themselves. As *Time* magazine reported, "... Choosing the sites for demonstration dumps will be a gigantic headache. No congressman's constituents will want to live next door to one."[19]

Indications of parochialism can be clearly seen in the major Senate nuclear waste proposal (S 1662). The core of the legislation is a timetable for research, development and decision-making, leading to the operation of a permanent high-level waste disposal site. However, some senators whose states have been prominently mentioned as possible disposal sites are call-

[19] *Time*, Oct. 26, 1981, p. 18.

ing for alternatives to a permanent waste repository. So the bill also includes provisions for a short-term, away-from-reactor (AFR) spent fuel storage facility, a long-term, monitored retrievable surface storage (MRSS) facility, and test and evaluation facilities at proposed permanent repository sites. According to Caroline LeGette of Congress Watch, the bill sets up "every kind of storage and disposal known, with the possible exception of 'Baggies.' "

The situation is further complicated by a jurisdictional dispute over defense-related radioactive waste. There is no question that the millions of gallons of high-level liquid waste that is temporarily stored in tanks in Washington state, South Carolina and Idaho will eventually have to be disposed of. The catch is that the defense establishment and its allies do not want any civilian regulatory authority to control the disposal of defense nuclear waste.

The Armed Services committees of both houses have jurisdiction over all nuclear defense programs, including the maintenance and disposal of defense nuclear wastes. Since the Nuclear Regulatory Commission will have supervisory and licensing authority over any civilian waste disposal facility, the Armed Services committees vehemently oppose the inclusion of defense waste in bills prepared by other committees that deal with nuclear waste.[20] They believe that a universal waste disposal plan would set an unwanted precedent of civilian control over defense programs, which could damage national security.

Opponents of the policy of waste separation are equally adamant. They say that defense-related wastes are just as hazardous as civilian wastes and should be subject to strict licensing and supervision by the Nuclear Regulatory Commission. "Separating the efforts to provide permanent disposal for defense-related nuclear waste will offer nothing to speed their disposal, but will in fact hamper, and possibly prevent, this nation's attempts to permanently isolate all such wastes from our environment," South Carolina Gov. Richard Riley (D) told the Senate Environment and Public Works Subcommittee on Nuclear Regulation. Supporters of combined, licensed facilities for civilian and military nuclear wastes also argue that separate facilities would be cost-inefficient.

The debate over the destiny of defense-related wastes was the main reason that high-level nuclear waste legislation failed to pass last December in the 96th Congress. Though both houses had passed legislation, the House bill included military wastes

---

[20] These committees include the Senate Environment and Public Works Committee, the Senate Energy Committee, the House Interior Committee, the House Science and Technology Committee and the House Energy and Commerce Committee.

in a disposal plan, while the Senate bill specifically excluded defense wastes. The two houses were able to compromise on other disparities between the two bills, but no agreement was ever reached on the defense waste problem, and the legislation died.[21]

Current Status of Waste Disposal Bills

The 97th Congress is trying to construct a workable nuclear waste disposal plan. In the Senate, the Energy and Natural Resources Committee and the Environment and Public Works Committee have approved different versions of a bill (S 1662) that contains authorization for development of high-level waste repositories, away-from-reactor (AFR) and monitored retrievable surface storage (MRSS) facilities, and test and evaluation projects. The two versions have some things in common. They both call for maximum state participation in site determination, give states a "veto" that could be overridden only by a majority of at least one house of Congress, and outline plans to tax utilities to create a fund to pay for the expensive disposal operations.

There are some differences, however, because the Environment Committee, which acted on the bill after the Energy Committee, added several amendments. One change would push the date for final NRC approval of a high-level waste repository site back from 1988 to 1991. Department of Energy officials and environmentalists had lobbied for the delay, arguing that the earlier date provided insufficient time to develop a definitive body of research.

The Environment Committee's other major alteration, which is likely to cause more controversy than the timetable change, deals with the subject of military nuclear waste disposal. The amendment calls on President Reagan to produce a plan by Jan. 1, 1983, for the permanent disposal of military waste. If the president cannot prove an overriding national security interest in separating civilian and defense nuclear wastes, any proposed defense waste site would be licensed and supervised by the Nuclear Regulatory Commission. This section is sure to result in intense debate, since it is almost certain that powerful members of the Senate Armed Services Committee, such as Chairman John Tower, R-Texas, and Henry Jackson, D-Wash., will attempt to remove any language that calls for licensing of military waste facilities.

In the House, the Interior Committee is considering a bill (HR 3809) that was reported by its Energy and Environment Subcommittee on Oct. 30. Far simpler than the Senate bill **it**

---

[21] See *1980 CQ Almanac*, pp. 494-502.

provides a timetable leading to NRC approval of a high-level waste repository site by 1990. When asked when the repository would be ready for operation, Interior Committee Chairman Morris K. Udall said, "No later than 1993." The House bill contains provisions for a state veto and a tax on utilities to fund the repository. It does not include language dealing with the military waste problem. The original text had contained defense waste provisions, but they were dropped because of opposition from the House Armed Services Committee.

On Nov. 20, the House Science Committee reported a bill (HR 5016) that includes an accelerated timetable for selection of a waste repository site, which would take place within three years of enactment. The bill also provides for development of an unlicensed test and evaluation facility. Like the bill reported by the Interior subcommittee, HR 5016 begs the question of military waste.

House Science Committee Chairwoman Marilyn Lloyd Bouquard, D-Tenn., has expressed interest in developing a consensus House bill on high-level nuclear waste disposal. However, no consensus will be reached until the third House committee with responsibility for nuclear wastes, the Energy and Commerce Committee, acts upon the question. This committee, which is chaired by Rep. John Dingell, D-Mich., has yet to schedule hearings on nuclear waste legislation, to the dismay of some of those involved in the waste controversy. According to George Gleason of the American Nuclear Energy Council, legislation is attainable "if the Energy and Commerce Committee will get off their butts and get moving."[22]

Even if Congress gets its act together, it does not mean that the nuclear waste issue is going to go away. There are too many unknowns involved — the future of reprocessing, the needs and demands of the military, the health of the nuclear power industry, the objections of residents in potential site areas, the possibility of a state veto, and the unsolved technological problems in geologic site selection — to state with confidence that Congress has the permanent solution to the high-level waste problem at hand. The nuclear waste issue may be as difficult to dispose of as the wastes themselves.

---

[22] Quoted in *The Washington Post*, Nov. 17, 1981.

Selected Bibliography

Books

Deese, David A., *Nuclear Power and Radioactive Waste*, Lexington, 1978.

Gilmore, W. R., ed. *Radioactive Waste Disposal: High and Low Level*, Noyes, 1978.

Shapiro, Fred C., *Radwaste: A Reporter's Investigation of a Growing Nuclear Menace*, Random House, 1981.

Willrich, Mason and Richard K. Lester, *Radioactive Waste: Management and Regulation*, Free Press, 1977.

Articles

Faltermeyer, Edmund, "Burying Nuclear Trash Where It Will Stay Put," *Fortune*, March 26, 1979.

Holton, Woody, "Congress Eyes Non-Answers for Nuclear Fuel," *The Congress Watcher*, October-November 1981.

Kerr, Richard A., "Geological Disposal of Nuclear Wastes: Salt's Lead is Challenged," *Science*, May 11, 1979.

Plattner, Andy, "Armed Services Committees Insist Military Be Exempted from Nuclear Waste Bills," *Congressional Quarterly Weekly Report*, Nov. 7, 1981.

—— "Nuclear Waste Legislation Begins to Move in Congress," *Congressional Quarterly Weekly Report*, Oct. 10, 1981.

—— "Reagan Nuclear Policy Draws Mixed Reaction in Congress," *Congressional Quarterly Weekly Report*, Oct. 17, 1981.

Shapiro, Fred C., "Nuclear Waste," *The New Yorker*, Oct. 19, 1981.

Smith, R. Jeffrey, "U.S. Urged to Reprocess Nuclear Fuel," *Science*, June 20, 1980.

Reports and Studies

Atomic Industrial Forum, "Statement of Position on the Storage and Disposal of Nuclear Waste in the Matter of Waste Confidence Rulemaking of the U.S. Nuclear Regulatory Commission," July 7, 1980.

Editorial Research Reports: "Nuclear Waste Disposal," 1976 Vol. II, p. 885.

General Accounting Office, "Is Spent Fuel or Waste From Reprocessed Spent Fuel Simpler to Dispose of?" June 12, 1981.

Interagency Review Group, "Report to the President on Nuclear Waste Management," March 1979.

Nuclear Regulatory Commission, "Regulation of Federal Radioactive Waste Activities," September 1979.

WESTERN OIL BOOM

by

Tom Arrandale

**May 29
1 9 8 1**

Editor's Note: The search for America's remaining oil and gas reserves is expected to accelerate through the 1980s. But a world oil glut, and political uncertainties in mid-1982, threatened to temper the U.S. petroleum industry's domestic drilling boom. Oil company planners were rethinking expensive drilling programs as the price of crude slid below $30 a barrel. Secretary of the Interior James G. Watt pushed ahead with plans to speed federal offshore oil and gas leasing. But under conservationist and congressional pressure, Watt backed away from opening Western wilderness areas to oil and gas exploration, at least until the end of the century. Meanwhile, the Reagan administration put on the back burner proposals to speed the lifting of natural gas price controls.

WESTERN OIL BOOM

THE U.S. oil and gas industry is going all out to find what's left of the nation's petroleum resources. At remote sites all over the country, giant rigs are drilling two miles or more into the earth in search of new deposits. In Alaska, along the Rocky Mountain chain and in offshore coastal waters, drilling crews are exploring what may be the last frontiers of America's petroleum era.

President Reagan's energy policies have spurred a drilling boom that has been building up across the continent since world oil prices began skyrocketing eight years ago. The president on Jan. 28 lifted the last federal oil price controls, and his administration has pledged to turn the hard-driving oil industry loose to find new onshore and offshore reserves.[1]

U.S. oil discoveries peaked half a century ago. Natural gas finds have fallen off since about 1950. At best, some experts maintain, stepped-up drilling will only briefly halt a steady fall in American oil and gas production and reserves. A recent Rand Corp. study conducted for the government suggests that "the U.S. petroleum industry is gradually running out of ideas as to where oil and gas may still be found."[2]

Western 'Overthrust Belt' Discoveries

But oilmen by nature are an optimistic breed. Many insist that new exploration techniques and deep drilling into unexplored formations will uncover deposits that were undetected or bypassed earlier. That hopeful view has proved out, at least in one long-neglected Rocky Mountain region. Since 1975, wildcat drilling[3] in Wyoming, Utah and Idaho has pierced 16 major oil and gas fields in previously unproductive terrain along a complex formation that geologists call the Western Overthrust Belt.

---

[1] See "Energy Policy: The New Administration," E.R.R., 1981 Vol. I, pp. 57-80.

[2] Richard Nehring, with E. Reginald Van Driest II, *The Discovery of Significant Oil and Gas Fields in the United States*, The Rand Corp., January 1981. The study was commissioned by the Department of Energy and the U.S. Geological Survey.

[3] In the oil industry, highly speculative drilling, especially beyond known deposits, is "wildcatting" or "wildcat drilling." Typically, this drilling is likely to be done by small independent companies or individuals.

In oilman's parlance, the thrust belt region is rapidly becoming known as "elephant country" — it holds gigantic oil and gas reserves. The Overthrust Belt was formed more than 100 million years ago when shifting subterranean land masses collided, pushing older, hard rocks over younger, softer formations. The softer formations often contain oil and gas. The Overthrust Belt runs 2,300 miles along the Rocky Mountain chain from northwestern Montana to southern Arizona. So far, at least, the oil and gas "play" — drilling activity — has been concentrated in southwestern Wyoming and northwestern Utah, roughly along the Union Pacific Railroad's right-of-way.

Evanston, Wyo., at the center of today's activity, was a stop on the Overland Trail in the mid-19th century — a place where the westward migrants greased their wagon axles with oil that seeped to the surface. Oil has been produced from fields along the edges of the thrust belt since the early 1900s. But drilling in the region during the 1940s and 1950s produced only dry holes, 200 in all.

In 1969 the Union Pacific reached agreement with Amoco Production Co., the domestic exploration subsidiary of Standard Oil of Indiana, to explore the railroad's land grants along a 40-mile belt from Colorado into Utah. That strip ran right across the Overthrust. American Quasar Petroleum, an independent from Fort Worth, Texas, working on a "farm out" arrangement with Amoco, set off the Overthrust boom in 1975. Drilling a 14,000-foot well, it struck oil and gas and thus discovered the Pineview field 40 miles east of Salt Lake City, Utah.[4]

That find "was an absolute shock to a lot of people," recalled Richard B. Powers, project chief for the Western Overthrust Belt at the U.S. Geological Survey Oil and Gas Section in Denver. Since then, drilling by Amoco, American Quasar, Chevron, Exxon, Energetics Inc. and other major and independent companies brought in 16 large fields — including six "giants" and one "supergiant" within a 100-mile-or-so stretch along the Idaho-Wyoming-Utah thrust belt.[5]

Comparison to Alaska and East Texas

In April, Amoco Production Co. President Leland C. Adams told New York financial analysts that the company's East Anschutz Ranch natural gas field on the Wyoming-Utah border could be "one of the most significant finds in North America

[4] For background see *The Overthrust Belt*, published by Petroleum Information Corp., Denver, Colo., 1978.
[5] As classified by the American Association of Petroleum Geologists, a giant field holds 100 million barrels of oil or 1 trillion cubic feet of natural gas. A supergiant field holds 500 million barrels or 3 — not 5 — trillion feet of gas.

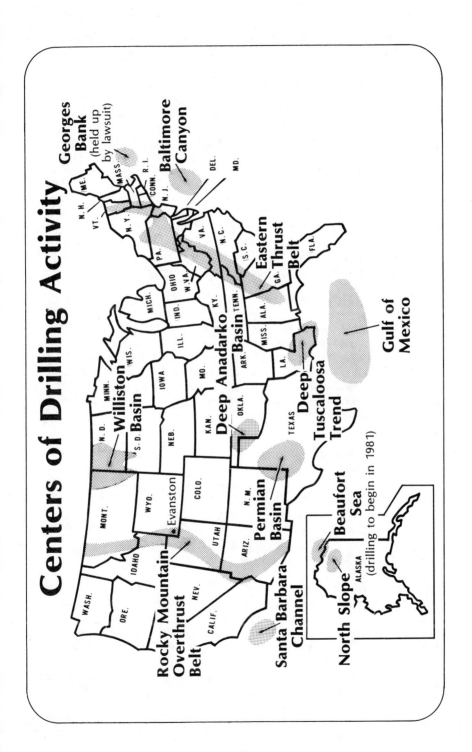

Centers of Drilling Activity

Georges Bank (held up by lawsuit)

Baltimore Canyon

Eastern Thrust Belt

Gulf of Mexico

Williston Basin

Deep Anadarko Basin

Deep Tuscaloosa Trend

Permian Basin

Rocky Mountain Overthrust Belt

Evanston

Santa Barbara Channel

Beaufort Sea

North Slope (drilling to begin in 1981)

ALASKA

WASH., ORE., CALIF., NEV., IDAHO, MONT., WYO., UTAH, ARIZ., COLO., N.M., TEXAS, N.D., S.D., NEB., KAN., OKLA., MINN., IOWA, MO., ARK., LA., WIS., ILL., IND., KY., TENN., MISS., ALA., MICH., OHIO, W.VA., VA., N.C., S.C., GA., FLA., PA., N.Y., VT., N.H., ME., MASS., R.I., CONN., N.J., DEL., MD.

since Prudhoe Bay was discovered" on Alaska's North Slope.[6] Powers estimated that reserves in the 16 fields amount to 3.2 billion barrels of oil and 16.5 trillion cubic feet of gas. "Every structure they're drilling out there seems to be of great size," Powers added in a May 12 telephone interview, "and they all seem to be filled right up to the spill point." Wyoming's Whitney Canyon-Carter Creek field, now being developed by Amoco, Chevron, Gulf and Champlin companies — the last, a Union Pacific subsidiary — "could certainly compare to East Texas," Powers said.

The 1930 East Texas discovery was one of the biggest ever in this country. While its production flooded the market in the 1930s, the East Texas field's entire reserves of five billion barrels are equivalent today to about the amount of oil America burns in one year. The great Prudhoe Bay reserves (9.6 billion barrels) fall slightly short of a two-year supply.

The size of the Overthrust Belt finds has forced geologists to keep lifting estimates of recoverable U.S. reserves. It also has brought energy company executives flocking to the Rocky Mountain region to set up offices in Denver and rigs on remote stretches of sagebrush country previously inhabited mainly by antelopes and wild horses. Last fall, 45 rigs were drilling in the Wyoming-Utah-Idaho thrust belt, probing as deep as 16,000 feet at costs that sometimes approached $7.5 million a well. "Everything about operating out there tends to be more expensive," said Owen Murphy, a Chevron spokesman in Denver.

Oilmen contend that crude oil prices now justify the added expense of exploring with computer-linked gear through rugged, wind-swept country and then drilling through hard and tortured rock formations. As a result, energy companies now are searching the far ends of the Overthrust Belt, from the mountain wilderness of western Montana to the saguaro deserts of southern Arizona. "So far, we've only scratched the surface," Amoco Vice President James Vanderbeek has said.[7]

Wide Range of Oil and Gas Drilling

The Overthrust Belt discoveries, along with rising crude oil prices, whetted the industry's appetite for exploring other regions. By January, 3,386 rotary drilling rigs were probing the nation's land and coastal waters, a 31.7 percent increase over the number at work the year before. In the same month, nearly 600 crews were using sophisticated electronic gear for seismic studies of deep and complicated formations with oil and gas

---

[6] See "Amoco: East Anschutz gas field is a big one," *Oil and Gas Journal*, April 6, 1981. p. 55.
[7] See "Rocky Mountain High," *Time*, Dec. 15, 1980, p. 28.

Wyoming-Utah-Idaho Thrust Belt

(Estimated Undiscovered Recoverable Reserves)

	High	Mean	Low
Oil *(billion barrels)*	13.3	6.7	2.7
Gas *(trillion cubic feet)*	105.4	58.4	29.1

Source: U.S. Geological Survey

potential. "There are some highly promising areas," said geologist Powers. Most new finds, he added, now result from "the rejuvenation of areas which had been picked over in the past, or just overlooked before."[8]

Drilling now is going on in geologic areas all over the country, from the Alaskan frontiers opened barely a decade ago to the Pennsylvania fields where former railroad conductor Edwin L. Drake drilled the first oil well in 1859. In Alaska, the Atlantic-Richfield Co.'s 1968 Prudhoe Bay discovery opened a "supergiant" field with 10 billion barrels of recoverable crude, the largest oil accumulation ever found in North America.[9] Alaska now produces 5,000 barrels of crude a day but remains, as described by the Rand Corp. study, "the last great frontier for petroleum exploration in the United States." Shell Oil Co. officials estimate that 60 percent of the petroleum still to be found in the country is in Alaska.

After the Prudhoe Bay find, exploration of many prospective Alaskan fields was delayed while Congress struggled to decide how much of the state's federally owned lands should be set aside for protection in wilderness and national parks.[10] In 1980, Congress finally agreed on legislation that doubled the size of the state's national parks and wildlife refuges and tripled protected wilderness areas.

In signing the measure into law on Dec. 2, 1980, President Carter contended that the Alaska lands measure nonetheless kept all offshore areas and 95 percent of the potentially productive oil and mineral areas open for exploration and drilling. But Congress designated as wilderness 80 percent of the vast William O. Douglas Wildlife Refuge, which geologists consider the most attractive potential oil and gas field left on the continent.

The government itself has been exploring the National Petroleum Reserve in Alaska. Oil companies have paid $1 billion to

---

[8] Telephone interview, May 12, 1981.
[9] See "Alaska Oil Boom," *E.R.R.*, 1969 Vol. II, pp. 835-854.
[10] See "Alaskan Development," *E.R.R.*, 1976 Vol. II, pp. 927-950.

obtain federal leases for exploring the Beaufort Basin off Alaska's icy northern coast, and the U.S. Geological Survey in January announced the discovery of five huge undersea basins off western Alaska in the Bering and Chukchi seas that could have enormous oil and gas resources.

Drilling is also intensifying in other areas of the nation. Companies are drilling and developing deep gas deposits in the southern Louisiana Tuscaloosa Trend and Oklahoma's Andarko Basin. On the northern Great Plains, about 150 rigs are drilling thin petroleum deposits under the Williston Basin, an empty region sprawling across North and South Dakota into eastern Montana. Only nine rigs were operating there as recently as 1972.

In the East, major and independent companies are leasing drilling rights and running seismic surveys along the 1,100-mile Eastern Overthrust Belt from upstate New York to Alabama. Meanwhile, producers are using advanced recovery methods to squeeze previously unrecoverable oil from long-producing fields in West Texas and California.

Off the nation's coastlines, interest remains high in potential deposits under the outer continental shelf[11] despite disappointing results from federally leased drilling in recent years in the Gulf of Alaska, eastern Gulf of Mexico, California's Outer Banks and Baltimore Canyon off the Atlantic coast.[12] Secretary of the Interior James G. Watt on April 12 announced a five-year offshore leasing program to step up exploration of promising regions under federal jurisdiction three miles or more offshore. Despite opposition from California Gov. Edmund G. Brown Jr. and environmental groups, Watt scheduled a sale of leases May 28 for drilling rights off the coast of Santa Barbara, Calif., where an oil well leak raised a strong protest in 1969 and became a rallying cry of the fledgling environmental movement.

James G. Watt
Secretary of the Interior

---

[11] The continental shelf refers to underwater land formations extending outward for varying distances from the shoreline. By one commonly used definition, it is the land under the offshore water out to a depth of 200 meters (660 feet). The *outer* continental shelf usually refers to waters of about that depth and, sometimes, still deeper. As legally defined for one specific purpose by the Outer Continental Shelf Act of 1953, the outer continental shelf is considered to begin three miles from the coast except in Texas and Florida, where it starts at three marine leagues offshore.

[12] See "Offshore Oil Search," *E.R.R.*, 1973 Vol. II, pp. 537-556.

Declining Reserves and Output

EVEN before the current drilling boom, the United States was the world's most extensively explored petroleum-producing country. Since Drake's 1859 discovery of "rock oil" near Titusville, Pa., wells have pumped roughly 100 billion barrels of oil from the 48 contiguous states and immediate offshore areas. Between 1859 and 1969, consulting geologist E. N. Tiratsoo wrote in 1973, the United States "produced more than 40 percent of all the crude oil produced anywhere in the world."[13]

The Northeast was the predominant oil-producing region in the 19th century, following the Pennsylvania discoveries. Oil was found in Colorado in 1862, but until 1900 exploration was largely limited to the Appalachian region and other areas like California and the Rocky Mountains where surface features suggested petroleum was present. But discoveries accelerated as geophysical techniques were perfected during the 1930s, and U.S. production shifted steadily south and west to California, Louisiana, Wyoming, Oklahoma, East Texas, and the Permian Basis of West Texas and eastern New Mexico.[14]

Before World War I, U.S. production accounted for between 50 and 90 percent of the world's oil. Even then, however, shortages were predicted, and the federal government set aside Naval Petroleum Reserves in 1912 to conserve fuel for oil-burning battleships. In 1925, the American Petroleum Institute (API) forecast "no imminent danger of the exhaustion of the petroleum reserves of the United States," but its report allowed for development of oil shale and assumed oil conservation efforts including improved gasoline engine efficiency.[15]

But the 1930 East Texas field discovery flooded the oil markets, and led several states to impose regulatory controls on production. In 1959, the federal government imposed quotas on cheap oil imports from the Middle East to preserve incentives for domestic production.

But U.S. crude oil production peaked at 9.6 million barrels a day in 1970. It dropped to 8.1 million barrels daily in 1976, then leveled off at 8.5 million barrels in 1981 as Alaskan output picked up. But the nation in the meantime grew dependent on imported crude for close to half its oil consumption, and the

---

[13] E. N. Tiratsoo, *Oilfields of the World* (1973).

[14] See Michel T. Halbouty, ed., *Giant Oil and Gas Fields of the United States* (1970).

[15] See H. William Menard, "Toward a Rational Strategy for Oil Exploration," *Scientific American*, January 1981, p. 55. For background on the long efforts to extract oil from rock shale in Colorado, Utah and Wyoming, see "Oil Shale Development," *E.R.R.*, 1968 Vol. II, pp. 903-922.

Organization of Petroleum Exporting Countries (OPEC) took advantage of the industrial world's need for Middle Eastern oil to jack prices from $3 a barrel in 1973 to well over $30 a barrel today.

Federal Energy Policies After 1973

Since the 1973 Arab oil embargo, the federal government has searched desperately for other energy sources. Under President Carter, the government stressed conservation measures to reduce foreign oil dependence and set up a Synthetic Fuel Corp. with a $20 billion congressional authorization to speed development of replacement fuels from domestic coal and oil shale. But President Reagan has a different view. In his acceptance speech at the Republican convention last summer, he said the nation "must get to work producing more energy," including conventional oil and gas.

The Reagan energy policies are still being formed. But the administration seems likely to scale back costly federal support for unproven synthetic fuel projects while encouraging domestic petroleum production. As a first step, the president moved quickly to "further stimulate domestic energy production and conservation" by ending oil price controls before their scheduled Sept. 30 termination date.

President Nixon first imposed domestic oil price controls in 1971 as part of an overall wage-price freeze. During the mid-1970s, a Democratic Congress blocked President Ford's proposal to lift the controls; they kept prices on 60 percent of U.S. production below rapidly climbing world levels. Ford argued that higher prices would encourage domestic oil development while curbing national oil consumption, but Democrats feared decontrol would feed inflationary pressures while enlarging the already considerable profits of U.S. oil companies.[16]

In April 1979, with Congress weakening his conservation program and gasoline shortages cropping up in some parts of the country, Carter announced he would phase out oil price controls under existing presidential powers. Nearly a year later, Congress imposed a "windfall profits tax" on oil companies, as Carter requested, to keep them from reaping "huge and undeserved profits" from decontrolled prices. The levy, actually an excise tax on added oil company revenues, was intended to raise $227 billion for the government from an estimated $1 trillion in additional industry income.

Carter's decontrol decision contributed to record oil industry profits in 1980. They were 32 percent above the previous year's

---

[16] For background, see Congressional Quarterly, *Congress and the Nation* Vol. IV, p. 242.

American Oil and Gas Resources

Mean 1981 Estimates of Undiscovered Reserves

	Oil (billion barrels)	Gas (trillion cubic feet)
Onshore Regions		
Alaska	6.9	36.6
Pacific Coast	4.4	14.7
Colorado Plateau and Basin and Range	14.2	90.1
Rocky Mountains and Northern Great Plains	9.4	45.8
West Texas and Eastern New Mexico	5.4	42.8
Gulf Coast	7.1	124.4
Mid-continent	4.4	44.5
Michigan Basin	1.1	5.1
Eastern Interior	0.9	2.7
Appalachians	0.6	20.1
Atlantic Coast	0.3	0.1
Total Onshore	**54.6**	**426.9**
Offshore Regions		
Alaska	12.3	64.6
Pacific Coast	3.8	6.9
Gulf of Mexico	6.5	71.9
Atlantic Coast	5.4	23.6
Total Offshore	**28.0**	**167.0**
TOTAL U.S.	**82.6**	**593.9**

Figures may not add to totals because of rounding.
Source: U.S. Geological Survey

profits. For 1980, sales by Exxon, Mobil, Texaco and Standard Oil of California ranked those four companies in the top five of the *Fortune* directory of the nation's 500 largest industrial corporations. Thirteen of the top 20 corporations listed by the magazine were oil companies.[17]

Oil profits fell in the first quarter of 1981, partly because

[17] "The 500," *Fortune*, May 4, 1981, p. 322.

rising gasoline prices curbed American driving habits. Saudi Arabia, meanwhile, kept its crude production at more than 10 million barrels a day and cut some prices by $2 a barrel in an effort to force other OPEC members to agree on unified price levels at their May 25 meeting. Oil was in surplus, at least temporarily, and major U.S. refiners cut gasoline prices, canceled contracts with OPEC producers and slashed the price they were willing to pay for domestic crude by $2 a barrel.

Still, oilmen credited decontrolled domestic prices for the stepped-up U.S. exploration and production. In February, an *Oil and Gas Journal* survey found that U.S. oil companies planned to spend $30 billion in 1981 on drilling and exploration in the United States, a 19.4 percent jump over 1980 commitments. Rotan-Mosle Inc., a Dallas financial firm, predicted early in 1981 that climbing oil and gas revenues will prompt the industry to put 5,700 rigs to work in the country by 1986.[18]

Uncertainty Over Reserve Estimates

How much petroleum is left for those rigs to find can only be determined by drilling. Oilmen insist that higher prices give them incentives to punch more wells into deep formations or in long-ignored regions too remote from proven reserves to justify drilling when prices were lower. J. D. Langston, vice president of exploration for Exxon, noted, "Something you'd call a dry hole some years back would now be called a discovery."[19]

Yet even as world crude prices climbed during the 1970s, the American Petroleum Institute's estimates of proven U.S. oil and gas reserves fell steadily despite a doubling of drilling activity. And estimates of undiscovered reserves remain largely a matter of guesswork. Extractive industries traditionally have been secretive about the reserves they hope to discover, and government and independent resource estimates usually vary widely.

The U.S. Geological Service on Feb. 25 updated its 1975 estimates of undiscovered oil and gas reserves. USGS geologists arrived at a mean estimate — half of the estimates were higher and half lower — of 82.6 billion barrels for undiscovered oil resources. This was only slightly above the 1975 figure, even after taking Overthrust Belt discoveries into account. The survey lifted its mean natural gas estimate 22 percent, to 594 trillion cubic feet, attributing much of the increase to offshore prospects in waters up to 2,500 meters (8,125 feet) deep. The 1975 study included offshore regions only to a depth of 200 meters.

[18] See *Oil and Gas Journal*, Feb. 2, 1981, p. 30; Feb. 16, 1981, p. 57.
[19] Quoted in *Newsweek*, Sept. 29, 1980, p. 67.

The study, widely used by government agencies and some oil companies, reduced 1975 estimates for offshore regions in the Gulf of Alaska, off Southern California and the eastern Gulf of Mexico, where exploratory drilling had been disappointing. But some industry experts believed the government's estimates overstated the likely resources left in aging fields such as the Permian Basin of West Texas. USGS and industry geologists found greater agreement on estimates for less-explored frontier regions like Alaska and offshore regions. But the widely varying reserve estimates raise questions "about the role of enthusiasm for the hunt," Richard A. Kerr suggested in *Science* magazine.[20]

Other assessments, based on historic patterns of petroleum discoveries, come up with less optimistic figures. For instance, the 1981 Rand Corp. report set onshore reserves left in the 48 contiguous states at one-tenth of the USGS estimate of 48 billion barrels. The study maintained that the bulk of the petroleum resources lay in giant oil and gas fields that had been found by mid-century. It argued that major new discoveries were unlikely "because of the increasing exhaustion of the geological possibilities region by region." Richard Nehring, the Rand report's author, told *U.S. News & World Report* that "we've found a lot of oil and gas in this country, but we've already produced most of it."[21]

Extending the Petroleum Era

OILMEN disagree with Nehring's conclusion, and this year they will sink 70,000 wells to step up production and enlarge U.S. petroleum reserves. High prices now justify drilling deeper wells in and around existing fields and tapping smaller known deposits. And Overthrust discoveries have borne out predictions that huge reserves lie in complex and little-understood traps that geologists once overlooked. "The potential in the United States is not only vast but promising," said Houston geologist Michael T. Halbouty, who led Reagan's energy advisory team during the post-election transition. And he added: "not only for the next 20 years but well into the 21st century."

But to extend the petroleum era beyond the turn of the century, the U.S. industry probably must find new giant deposits in such frontier regions as Alaska and offshore waters. Ex-

---

[20] Richard A. Kerr, "How Much Oil, It Depends on Whom You Ask," *Science*, April 24, 1981, p. 427.
[21] Quoted in *U.S. News & World Report*, May 4, 1981, p. 49.

cept along the western Overthrust Belt, few large fields are likely to be found in extensively drilled regions; and smaller fields probably cannot be emptied fast enough to meet short-term demand for liquid fuels. "Most of the oil that can be extracted quickly is in giant fields," H. William Menard argued in a *Scientific American* study of petroleum exploration strategies. "The only way for the U.S. to meet most or all of its requirements for oil over the next two decades is to discover a substantial number of giant fields, no fewer, say, than 100."[22]

As Menard pointed out, geophysical exploration has identified potential oil-bearing structures of gigantic size on federal lands in Alaska and the outer continental shelf. In addition, the Wyoming and Utah thrust belt discoveries are in formations that twist and plunge under federally owned parks, forests, rangelands and wilderness areas in remote, largely unspoiled country. As a result, the federal government eventually must come to grips with what Harvard Business School Professor Robert Stobaugh, co-editor of a book on the nation's *Energy Future,* has called "one of the toughest value judgments the nation has to make."[23]

Reagan's 'Unlocking' of Federal Lands

The U.S. Geological Survey has no precise estimates, but geologists generally believe that federal lands, including offshore deposits, hold half of the nation's undiscovered oil and two-fifths of the natural gas. The federal government has leased about 100 million acres of public lands and roughly 19 million on the outer continental shelf for oil and gas development. But throughout the 1970s, the industry charged that U.S. land management policies blocked exploration and development of promising regions.

A congressional General Accounting Office (GAO) study released in February 1981 calculated that federal agencies had closed 64 million acres, about 15 percent of all U.S. holdings, in 48 continental states to oil and gas leasing. Congress, in passing the 1980 Alaska lands law, restricted mineral development by varying degrees as it set aside 104.3 million acres in federal conservation units, including 56.7 million acres for wilderness preservation.[24] And during Carter's conservation-minded administration, the Interior Department's Bureau of Land Management and the Agriculture Department's Forest Service limited access to roadless lands that the agencies were studying

---

[22] H. William Menard, "Toward a Rational Strategy for Oil Exploration," *Scientific American,* January 1981, p. 55.
[23] Quoted by Rich Jaroslovsky in *The Wall Street Journal,* April 1, 1981.
[24] See Congressional Quarterly *Weekly Report,* Aug. 23, 1980, p. 2447; Nov. 15, 1980, p. 3377.

Natural gas well drilling in Wyoming's Overthrust Belt

as potential additions to the nation's wilderness sytem. As a result, Chevron spokesman Murphy complained, "there were an awful lot of places that were just plain locked up without sufficient justification."

Interior Secretary Watt, chairman of Reagan's Cabinet council on natural resource policy, has pledged to speed leasing of public lands previously off limits to development. "The lands that have energy and mineral potential can have those resources

explored — and developed and produced — without destroying
the other qualities," Watt has maintained. "To deny access to
those lands for development is to deny the realization of some of
their prime value."[25]

Environmental Disputes Over Leasing

BLM sells leases for oil and gas exploration on public lands,
including the national forests, while the U.S. Geological Survey
issues permits for actual drilling. The leasing process has grown
long and complicated in recent years as Congress enacted laws
to protect endangered species, preserve archeological ruins and
control environmental damage of federal lands. Even when
leases are issued, agencies often attach special conditions to
lessen disruption to wildlife and natural landscapes.

In some places, including at least 345,000 acres in the Wyoming thrust belt, leases bar any surface disturbance at all,
forcing operators to resort to slant drilling from adjacent lands
and other costly measures. "There's an awful lot of bureaucratic
red tape involved, the kind of things that make it difficult to
work on BLM lands," said Peter Hanagan, executive director of
the New Mexico Oil and Gas Association.

Watt plans to expedite the leasing process, and the administration already has moved to open areas that had been closed
until agencies studied them for possible wilderness protection.
Following a Forest Service study of roadless regions in national
forests, President Carter proposed that Congress expand the
wilderness system to 33 million acres, mostly in the West.

The Forest Service designed an additional 10.8 million acres
of roadless areas for further study of their wilderness potential.
Meanwhile, BLM has identified 23.8 million acres of western
public lands as wilderness study areas. In the past, at least, both
agencies have strictly limited any intrusions that might destroy
the wilderness before the government decided whether to preserve them.

Conservationists contend that most areas with the best oil
and gas potential are no longer being considered by Forest
Service and BLM. The GAO study, in assessing public lands in
six states, found that 16.5 million acres of potential wilderness
included "at least 8.5 million acres . . . prospectively valuable for
oil and gas." In Wyoming alone, the study noted, oil and gas
could lie under 483,000 acres of BLM wilderness study areas
and 1.3 million acres that the Forest Service recommended for
preservation. In the six states studied — Wyoming, Colorado,

---

[25] See "How Interior Plans to 'Unlock' Public Lands," *Business Week*, March 23, 1981, p.
84L.

New Mexico, Utah, Nevada and Mississippi — the GAO estimated that wilderness lands could hold 387.4 billion barrels of oil and 162.4 billion cubic feet of gas.[26]

Watt already has moved to make it easier for companies to develop existing leases on areas that BLM has identified for wilderness study — if drilling causes no "unnecessary or undue degradation" of the land. Watt reportedly also is weighing a proposal that Congress extend a deadline set by the Wilderness Act of 1964 that will prohibit mineral leasing in wilderness areas starting in 1984.

In setting up the wilderness system, Congress in effect gave the industry 20 years to lease and find minerals within protected areas. But company spokesmen argue that the Forest Service has administered wilderness areas in ways that precluded exploration, while government rules now block access to potential additions to the system. Alice Frell, public lands director for the Rocky Mountain Oil and Gas Association in Denver, said in a telephone interview that "areas being studied for wilderness should be explored so the people will know what is being given up."

Conservation groups plan to fight Watt's efforts to grant entry to unspoiled regions. In mid-May, the Forest Service turned down a geophysical company request to run seismic studies, including 5,400 dynamite blasts, in the Bob Marshall Wilderness in Montana, near Glacier National Park, after environmentalists objected that the activity might intrude on grizzly bear habitat and endanger the country's biggest herd of bighorn sheep. On May 21 the U.S. House Interior Committee used an emergency provision of federal law to prohibit petroleum exploration in the 1.5-million acre Bob Marshall Wilderness. The committee's action does not require further congressional approval but it may be subject to challenge in the courts. The 23-18 committee vote was along party lines, with Democrats in the majority.

In the Overthrust region, "it's just such a massive onslaught right now that we're losing everything," protested Bruce Hamilton of Lander, Wyo., the Sierra Club's Northern Plains representative. "Apparently now they'll go a couple of steps further and go after the designated wilderness areas and the national parks."

Exploratory drilling along Wyoming Overthrust Belt now is moving north into national forests around Jackson Hole, Wyo.,

---

[26] *Actions Needed to Increase Federal Onshore Oil and Gas Exploration and Development,* General Accounting Office, Feb. 11, 1981.

and Grand Teton National Park. Oil companies are waiting for a Forest Service decision on whether to permit drilling in the East Palisades region, a 250,000-square-mile area straddling the Idaho border that has been designated for further wilderness study.

"We aren't saying there shouldn't be any exploration whatsoever," Hamilton commented. "But there ought to be a way to do it that guarantees that the wilderness potential of the area won't be diminished." Conservationists argued that western wilderness lands hold at best enough petroleum to supply a few days worth of national energy consumption. Brant Calkin, the Sierra Club's Southwest representative, contends, "that's not adequate promise to put in roads that will be there a lot longer than the oil will."

Growing Energy Role for Natural Gas

Eventually, of course, the nation will run out of crude oil reserves, and stepped-up drilling and production may only exhaust that resource more quickly. Alternative fuels like coal, oil shale, nuclear and solar power all face a variety of economic or environmental obstacles. But large new discoveries, along with Reagan administration policies, are raising hopes that natural gas can bridge the gap between the decline of oil and the development of a replacement source of long-term energy. "Natural gas can last long enough to buy time for new energy sources," said Roby Clark, president of the American Association of Petroleum Geologists. Once flared off as an unwanted byproduct from oil well production, natural gas has developed as an important industrial and home-heating fuel. Pipelines built after World War II connected gas-producing areas of the South and Southwest with the industrial Northeast and Midwest. The nation now consumes 20 trillion cubic feet a year, and natural gas reserves have dropped to about a ten-year supply as use outpaces new discoveries.

But in a 1979 study, the American Gas Association estimated that between 500 trillion and 1,000 trillion cubic feet of gas remained in undiscovered U.S. reserves. Congress in 1978 approved a plan to deregulate natural gas prices, which had been controlled at the wellhead by federal regulations since 1954. The 1978 Natural Gas Policy Act, a compromise between congressional delegations from gas-producing and gas-consuming states, set up a complicated schedule for phasing out price controls on gas by 1985. But the 1978 law immediately deregulated gas found by new drilling in deep formations, 15,000 feet or more beneath the surface. That incentive lead to 30 percent more seismic exploration in 1979; and in 1980 the industry drilled nearly 16,000 gas wells, according to the American Petroleum Institute, including nearly 2,000 wildcat wells that tapped gas in new formations.

"They never drilled specifically for gas before," USGS geologist Powers noted. "It's deep gas they're looking for, and there are some huge deposits at 15,000 to 16,000 feet." Overthrust Belt discoveries produce more gas than oil, and the Eastern Overthrust, Tuscaloosa Trend and offshore Alaskan areas also have promising natural gas potential.[27]

The Reagan administration reportedly planned to encourage use of natural gas by asking Congress to repeal another 1978 law. It was enacted to force many utilities and industrial plants to switch from oil and gas to coal. "It is well established that there are significant new fields of gas that were literally unknown five years ago," National Gas Supply Association President David Foster contended in 1980. "We don't need to be so panicky about converting to coal."

Now that oil price controls have been completely lifted, the administration may also propose more rapid deregulation of natural gas. Since the 1978 gas law was written, world crude oil prices have risen so quickly that current prices more than double the price that Congress set for gas in the final year of regulation ($2.97 per thousand cubic feet). Industry observers argue that the relatively low price for gas continues to discourage more exploration. They also worry that Congress will put a windfall profits tax on gas if prices jump overnight when regulation expires as scheduled. Sen. James A. McClure, R-Idaho, chairman of the Senate Energy Committee, said early this year that the panel would study gas deregulation but predicted that Congress was unlikely to advance the 1985 expiration date.

Rising prices already have spurred more drilling, and access to federal lands and offshore leasing could open up new fields. But many experts doubt that exploration and development can more than briefly reverse the continued decline of U.S. oil and gas production. Stobaugh and co-editor Daniel Yergin, in their book's conclusion, suggested that "the United States will be fortunate if it finds enough new oil and gas to keep production at current levels."

Major oil companies themselves have begun to diversify in anticipation of the waning of the petroleum era. Some have moved into oil shale, coal, uranium, even department stores and office equipment. Some cash-rich corporations in 1981 began acquiring mining companies to develop other businesses. "The

---

[27] See "Gas Strikes Spark Play in Eastern Overthrust," *Oil and Gas Journal,* April 27, 1981, p. 109.

oil companies have an ample amount of money to spend," a western financial consultant said. "They could put it into drilling and producing oil," he added, "but face it, you can only drill so much."

Selected Bibliography

Books

Halbouty, Michel T., ed., *Geology of Giant Petroleum Fields,* American Association of Petroleum Geologists, 1970.

Petroleum Information Corp., *The Overthrust Belt,* 1978.

Stobaugh, Robert and Daniel Yergin, eds., *Energy Future, Report of the Energy Project at the Harvard Business School,* Random House, 1979.

Tiratsoo, E. N., *Oilfields of the World,* Scientific Press Ltd., England, 1973.

Articles

Holt, Donald D., "How Amoco Finds All That Oil," *Fortune,* Sept. 8, 1980.

"How Interior Plans to 'Unlock' Public Lands," *Business Week,* March 23, 1981.

Kerr, Richard A., "How Much Oil? It Depends on Whom You Ask," *Science,* April 24, 1981.

Menard, H. William, "Toward a Rational Strategy for Oil Exploration," *Scientific American,* January 1981.

"Oil Fever Rages — And So Does an Old Debate," *U.S. News & World Report,* May 4, 1981.

Oil and Gas Journal, selected issues.

"Rocky Mountain High," *Time,* Dec. 15, 1980.

Reports and Studies

Editorial Reseach Reports: "Offshore Oil Research," 1973 Vol. II, p. 539; "Western Land Policy," 1978 Vol. I, p. 83; "Oil Antitrust Action," 1978 Vol. I, p. 103; "Energy Policy: The New Administration," 1981 Vol. I, p. 59.

Nehring, Richard, with E. Reginald Van Driest II, "The Discovery of Significant Oil and Gas Fields in the United States," The Rand Corp., 1981.

Powers, Richard B., "Assessment of Oil and Gas Resources in the Idaho-Wyoming Thrust Belt," Wyoming Geological Association Guidebook, 1977.

Cover and inside illustrations by Staff Artist Robert Redding

SOLAR ENERGY'S UNEASY TRANSITION

by

Jean Rosenblatt

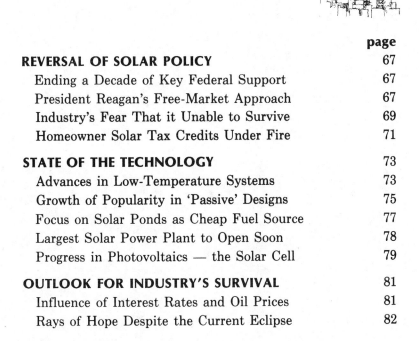

Mar. 26
1 9 8 2

SOLAR ENERGY'S UNEASY TRANSITION

S HOCKED by the energy crisis and awakened to environmental concerns, this nation began about a decade ago to rediscover such ancient sources of energy as sun, wind and water.[1] Congress for the first time in 1973 made money available for the development of solar energy, through the National Science Foundation. These funds grew steadily, peaking at $500 million in President Carter's budget for fiscal year 1981.

Carter made solar research and development an important part of his energy program, placing it in the new Department of Energy. The government arranged in 1977 to establish the Solar Energy Research Institute (SERI) near Golden, Colo., and participated in "Sun Day," May 3, 1978, to heighten public awareness of solar energy's potential as a national resource. Then in 1979 Carter committed the United States to a goal of meeting 20 percent of its energy needs with solar and other renewable resources by the year 2000.

Ronald Reagan, his successor, wasted no time in reversing those priorities. His administration instead emphasized petroleum and nuclear power.[2] In his first year as president, Reagan cut solar energy funding in about half and he proposes further cuts in his new budget *(see box, p. 72)*. This turnabout in policy has left the fledgling solar industry unsure, despite rapid technical advances, whether it can survive in the marketplace.

President Reagan's Free-Market Approach

"The free market will determine the development and introduction of rates of solar technologies consistent with their economic potential," Reagan said in the federal budget he sent Congress in January for the 1983 fiscal year, beginning this Oct. 1. The same level of federal support was no longer necessary, the Reagan administration reasoned, because more capital would be available for private investment in renewable energy sources as a result of the president's economic recovery program. Regulatory relief, rising energy prices and economic incentives such as tax credits for energy investment would also boost solar development in the absence of more direct forms of federal

---

[1] For background on the rethinking of the nation's proper long-term energy sources, see "New Energy Sources," *E.R.R.*, 1973 Vol. I, pp. 185-204.
[2] See "Energy Policy: The New Administration," *E.R.R.*, 1981 Vol. I, pp. 57-80.

support, according to the same argument. A recent Department
of Energy summary report to Congress stated:

> Despite technological progress and impressive percentage
> growth, renewable energy forms currently remain a small factor
> in the overall national picture (except for hydropower and the
> wood wastes that have long been used in the pulp and paper
> industry). Active solar, geothermal and wind systems together
> produced considerably less than 1 percent of the energy con-
> sumed in the United States last year. Their potential for the
> midterm to distant future is great, but the critical factors govern-
> ing their rates of growth are primarily economic. This recognition
> underlies [the] change in focus and objectives for most of the
> renewable energy programs.[3]

Only $2.2 billion of the administration's $11.8 billion proposed budget for the Department of Energy would be spent for
research and development. The bulk of it — $5.5 billion —
would go for the development and production of nuclear weapons.[4] The proposed $72 million solar budget would close out the
following programs: active and passive *(see Solar Terminology)*
systems development; ocean energy research and development;
information collection and dissemination. It also would delete
funds to build a permanent facility for the federal Solar Energy
Research Institute. The administration, proposing to abolish
the entire department, has asked that the remaining programs
be funded at the minimum level necessary for their transfer to
other agencies.

This energy policy has suffered a setback in Congress. On
March 3 the Senate Energy Committee overwhelmingly rejected
Reagan's proposals to dismantle the department and slash 1983
funding for solar research programs. The committee voted to
maintain spending for research on alternate energy sources,
including solar, at current levels. Solar advocates in Washington
predict the solar budget will be frozen at 1982 levels and that
the department is safe for the time being. But a budget even at
1982 levels is only "a victory in terms of the times," said Scott
Sklar, political director of Washington's Solar Lobby, in a recent interview.

Congressional misgivings about the new federal energy policy
seem to be shared by the public. A national poll conducted by
the Gallup organization for SERI at the time of the 1980 presidential election found that solar energy ranked first as the
preferred energy source and nuclear ranked last, behind conservation, synthetic fuels, water power, coal, and petroleum, in
that order.

---

[3] Department of Energy, "Sunset Review," Vol. 1, 1982, p. 26.
[4] This is done by the Department of Energy on behalf of the Department of Defense.

Sun Power

The sun is this planet's most abundant source of energy. Only an infinitesimal fraction of the sun's radiant energy strikes this planet, but our share still equals about 180 trillion kilowatts of electricity, more than 25,000 times the world's present industrial power capacity. The solar energy reaching the surface of the United States annually is greater than the total amount of fossil-fuel energy that, scientists say, will *ever* be extracted in this country.

This energy can be captured either directly through such devices as rooftop collectors, photovoltaic cells and building design features, or indirectly through the storage of solar energy in nature. The solar energy in trees, grasses, agricultural wastes, garbage and other organic materials can be burned to produce electricity or synthetic fuels. Even wind, which turns windmills to supply power, is an indirect form of solar energy. And so is water that has been heated by the sun to produce power.

Throughout the 1970s Congress tried to make the government a buffer against economic risks that accompanied developing, unproven solar technologies. Federal buildings, including military housing, provided a place for determining if solar technologies were commercially feasible. Federal tax credits and subsidized loans for residential and commercial solar applications also became available. In addition, the federal government made big investments in solar research and development, and in making solar information available to the public.[5] These elements provided a crucial boost to the solar industry in its formative years. Now, thousands of companies, individuals and citizen groups are experimenting with, buying, selling, installing and servicing solar energy systems.

Industry's Fear That it Unable to Survive

The Reagan administration's cuts in the federal solar budget and its de-emphasis of solar as a significant energy source have angered solar advocates, who feel that the nation's most promising energy future is being deserted. One outspoken critic of Reagan's energy policy is solar advocate Denis Hayes, former director of the Solar Energy Research Institute and an origina-

[5] See J. Glen Moore, "Solar Power," Congressional Research Service, Library of Congress, Feb. 18, 1982.

Solar Terminology

Renewable energy resources refer to energy sources that are virtually inexhaustible, such as the sun, wind, water, the Earth's heat (geothermal energy) and biomass (agricultural crops, aquatic plants and wastes and residues from agricultural products, wood and animals).

Solar energy is the energy transmitted from the sun in the form of electromagnetic radiation. Because the sun is an indirect source of most renewable energy, the terms solar and renewable energy are often used interchangeably.

Active solar systems need mechanical means such as motors, pumps or valves to operate.

Passive solar systems use structural rather than mechanical devices to collect and transfer solar energy.

Solar collectors capture and accumulate the sun's heat. The medium used to transfer the heat to other parts of the system varies.

Photovoltaics is the process by which energy from the sun is converted into electricity.

tor of Sun Day. Hayes was asked to resign last June after the institute's budget had been cut by more than half and its staff reduced from 850 to 580. In a speech to SERI employees at that time, Hayes said:

> Secretary of Energy James Edwards has embarked upon a careful, methodical campaign to destroy America's best energy hope. The shifts in the energy budget have been described by administration spokesmen as pure exercises to trim the federal [1982] budget. That is a manifest lie. If the budget were being trimmed for purely macroeconomic reasons, the nuclear budget would not be increased by 36 percent, while the solar budget was slashed by 67 percent, and the conservation budget was cut by more than three-fourths. . . . It is not an overdramatization to say that this administration — and, in particular, Secretary Edwards — has declared open war on solar energy.[6]

The central issue, experts agree, is whether the solar industry is strong enough to enter the market without continued federal help. During budget authorization hearings in February 1981, Edwards predicted that cuts in the solar budget would have little effect on the use of solar energy. Rep. Robert S. Walker, R-Pa., has defended the budget cuts as an effort to inject a new philosophy into solar development. "The prime inhibiting factors to solar energy have been economic," he said last summer. "They've been the high interest rates . . . the lack of capital investment money available. This administration through its policies of decontrol and deregulation, through . . . trying to

---

[6] Text printed in the Solar Lobby's magazine, *Sun Times*, July-August 1981, p. 10.

bring down inflation and interest rates, will do more to help the development of solar energy than the government subsidy program that the past administration relied on so heavily."

One who does not share Walker's view is Tony Adler, a New York investment banker specializing in solar energy. "We can easily lose our technological edge to the Japanese if the government does not make a commitment both in terms of financing and setting economic goals," he said.[7] Scott Sklar of the Solar Lobby predicted to Editorial Research Reports that "in the years when renewable energy comes to its height, we'll be importing most of the technology."

Certainly the impact of the new federal philosophy will be felt unevenly throughout the solar industry, which is made up of companies ranging in size from small manufacturers of flat-plate thermal collectors that heat water to oil companies working on complex photovoltaic development. Smaller companies that have been dependent on government research projects will have to do the research themselves. Many of them cannot do it and still keep their prices down. According to Paul Cronin, past president of the Solar Energy Industries Association (SEIA) in Washington, D.C., "research and development for future solar technologies, and those areas that are very capital intensive — I don't think industry has the funds that would be needed to follow these projects through to the end."[8]

Budget cuts have been a negative signal to consumers, investors and utilities, Carlo La Porta, SERI's director of marketing, research and technology, said in an interview. Utility companies, in particular, are at a critical stage in their investigation of solar and renewable energy sources. He said many are disillusioned with nuclear energy's potential because of high costs and safety factors. These corporations need demonstrations of solar technology at work — demonstrations that previously were funded by the government.

Homeowner Solar Tax Credits Under Fire

Chief among the solar industry's concerns is the fate of federal tax credits, which the House Science and Technology Committee estimates will be worth about $10 billion between now and 1986. "That's a major commitment toward the development of solar energy in the near term," according to Rep. Walker. However, last Sept. 24 President Reagan spoke on national television and discussed some tax policy changes the Treasury Department would consider to raise federal revenues. One of these changes involved eliminating "certain energy tax credits for businesses and individuals."

[7] Quoted in *The New York Times*, Aug. 19, 1981.
[8] Appearing on "The MacNeil-Lehrer Report," PBS-TV, July 7, 1981.

Funding for Solar Programs
(in millions of dollars)

	FY81	Reagan FY82 Request	Final FY82	Reagan FY83 Request
Active solar	41	12	12	0
Passive solar	32	10	11	0
Photovoltaics	139	63	74	27
Wind energy	60	19	34	5
Ocean energy	35	0	19	0
Industrial solar*	169	75	74	28
Other solar	24	15	24	12
Solar reserve account	0	9	15	9
Total	500	193	256	72

*Includes solar thermal, biomass and alcohol fuels
Figures may not add to totals because of rounding
Source: Congressional Research Service, Library of Congress

Congress enacted tax credits for businesses and individuals in 1978 as part of the Energy Tax Act, and supplemented them in the 1980 Windfall Profits Tax Act that accompanied the decontrol of domestic oil prices. Until the end of 1985, the expiration date under present law, a taxpayer who installs solar equipment in his new or existing home may claim a 40 percent credit on the first $10,000 he spends — thus reducing his federal income taxes by up to $4,000. Businesses that install solar equipment qualify for a 15 percent credit, in addition to the regular 10 percent business investment credit.

The solar industry was alarmed at Reagan's hint of seeking a repeal of energy tax credits. Immediately after his televised speech, the Solar Lobby, the Solar Energy Industries Association and other solar groups got together and presented their case in Congress. Resolutions defending the credits were introduced in the House and Senate, where they gained 263 and 64 cosponsors, respectively. The administration has never formally moved to repeal the energy credits, but solar advocates remain vigilant.

R. Nicholas Loope, SEIA president, considers tax credits "the single most important incentive for consumers to purchase solar equipment and for businesses to grow." Internal Revenue Service figures show that in 1978, the first full year of the residential tax credit program, 57,901 taxpayers reported spending $120.3 million on solar projects. The next year 61,245 returns listed $165.5 million in solar outlays, and in 1980 some $395 million worth of solar expenditures were claimed on 136,000 returns.

Forty-four state governments also provide tax breaks, including property and sales tax exemptions, income tax deductions and credits, for people who install solar systems.[9] In August 1981, Gov. Hugh L. Carey signed a bill making New York second only to California in providing the most liberal tax incentives to homeowners who buy solar systems.[10] The law provides a 55 percent tax credit, up to $2,750, for installing a rooftop solar collector or incorporating passive solar features into a building's design. New York also exempts solar equipment in determining a home's property tax assessment.

State of the Technology

SOLAR scientists generally divide solar energy systems into those that produce low, medium and high temperatures. Much of the recent work in low-temperature (100-150 degrees Celsius; 212-302 degrees Fahrenheit) active technologies has focused on reducing the manufacturing cost of solar collectors, improving their efficiency and reliability, and simplifying their design and installation. In spite of significant advances, however, the Department of Housing and Urban Development, which has subsidized more than 20,000 installations of active solar systems, reported last year that "system quality has been generally poor and in some cases it has been terrible." Owners have complained chiefly about shoddy installation and poor service. According to HUD and Energy officials, though, service is improving.

A flat-plate collector is typically the key component of an active low-temperature system. It is a rectangular, shallow box with a glass or transparent plastic lid and a dark bottom that absorbs incoming sunlight. The box traps the light and converts it into heat. Since 1975, collectors and other components have been designed for "very low temperatures" (under 45 degrees C; 100 F). Typically they heat swimming pools.

"While the cost of flat-plate collectors has increased at about the rate of general inflation, the value of the product is now much higher than it was five years ago," writes Melvin K. Simmons, a General Electric Co. scientist in Schenectady, N.Y.[11] Advances are being made in the methods and materials

---

[9] States without such incentives are Kentucky, Minnesota, Mississippi, Pennsylvania, West Virginia and Wyoming.
[10] According to SERI, California accounts for about half of the country's sales of solar water heaters.
[11] Melvin K. Simmons, "Solar Energy Technology — a Five-Year Update," *Annual Review of Energy*, 1981, p. 6.

for glazing the collector and absorber surfaces. Glass is still the preferred glazing in most cases, but now plastic and composite glazing is available. It is inexpensive and lightweight, an important factor in the installations of rooftop collectors. However, while efficiency of plastic and composite materials rivals that of glass when new, scientists are still concerned with the effects of long exposure to the environment, especially to ultraviolet light, a component of sunlight.

General Electric is testing a new type of plastic collector which its designers call "revolutionary" because it can turn sunlight into heat even on a cold, cloudy day. The unit would cost about a third as much as collectors now on the market — collectors are usually four-by-twelve feet and cost about $30 per square foot. And they would weigh only a little more than three pounds per square foot, considerably less than standard collectors. The current cost of standard collectors, even with federal and local tax incentives, deters many consumers, particularly those who live in Northern states where in winter collectors sometimes provide only 15 percent of the required space heating.

The cover of GE's new unit is made up of cylindrical vacuum tubes rather than flat glass or plastic. This design allows the collector to absorb more light in less area, according to John Flock, manager of GE's chemical engineering technical unit. "During cloudy days, the sunlight hits the cloud and becomes diffused," Flock said. "Unlike direct sunlight, more light is bounced off the clouds. The collector tubes, then, help to collect more energy at more angles." [12] GE's collector is not yet on the market because it is still being tested and because there are not enough consumers ready to buy, the company has concluded. The major market now is in hot water, not residential heating.

Despite complaints about shoddy installation, the active solar industry has grown rapidly, with the help of tax incentives and federally funded demonstration projects to prove the technology. Until 1974 there was virtually no market for solar equipment, but it is now being used in 300,000 to 400,000 buildings in the United States, according to the Department of Energy. The industry has grown from fewer than 50 collector manufacturers earning about $17 million in 1975 to a $400 million enterprise involving more than 300 companies. Of these, the 10 largest claim about three-quarters of the business. About 7,000 companies — many of which are heating and plumbing businesses with solar lines — are involved in installing active systems.[13]

---

[12] Quoted in *The Christian Science Monitor*, Feb. 24, 1982. See also "Solar Breakthrough Looks Bright at GE," *Industry Week*, Feb. 22, 1982.

[13] The last two estimates were made by the Solar Research Energy Institute in "New and Renewable Energy in the United States of America," a report prepared for the United Nations Conference on New and Renewable Sources of Energy held in Kenya last August.

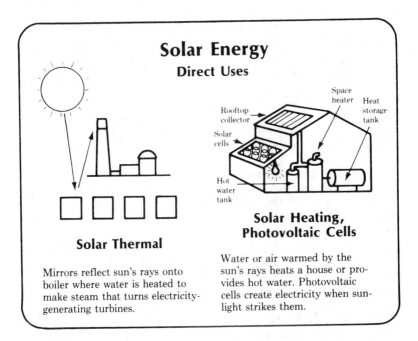

Solar Energy
Direct Uses

Space heater Heat storage tank

Rooftop collector

Solar cells

Hot water tank

Solar Thermal

Mirrors reflect sun's rays onto boiler where water is heated to make steam that turns electricity-generating turbines.

Solar Heating, Photovoltaic Cells

Water or air warmed by the sun's rays heats a house or provides hot water. Photovoltaic cells create electricity when sunlight strikes them.

According to the Department of Energy, fuel savings from solar energy systems installed so far are equivalent to between 1.0 and 1.5 million barrels of oil *annually.* This country's oil consumption currently is about 16 million barrels *daily.* "The overall effect is still minor," the department said in its recent report to Congress, "but generally it has been to increase employment."

This appraisal is a stark contrast to ecologist Lester R. Brown's assessment in his book *Building a Sustainable Society.* "With energy prices and public incentives increasing in the United States," Brown wrote, "rooftops may well sprout solar collectors in the Eighties the way they did television antennas in the Fifties.... As fossil fuel supplies dwindle ... and as the costs of solar collectors fall with mass production, solar collectors can be expected to meet a steadily expanding share of the world's need for low-temperature heat." [14]

Growth of Popularity in 'Passive' Designs

There is some indication that active technologies may be giving way in popularity to passive ones. In February, the consulting firm of Booz-Allen & Hamilton Inc. completed a study for the Department of Energy on 65 commercial buildings. All had been built during the last 10 years and had won awards for their energy-saving designs. The firm found a "mod-

---

[14] Lester R. Brown, *Building a Sustainable Society* (1981), pp. 232-233.

est shift" from active to passive designs. "It's a little premature to say that a trend is emerging, but this may be the first sign," said G. Kimball Hart of Booz-Allen's energy and environment division.[15]

An indication of the passive design's growing popularity is reflected in estimates by the Department of Energy that the number of passive solar homes increased from 500 in 1977 to 60,000 by the end of 1981, a time of decline in home construction. Most passive features, such as the site of the house and the arrangement of rooms and windows, must be built into a new house rather than added to an existing one. Experts attribute the popularity of passive solar housing partly to the loss of federal demonstration grants for active systems and partly to homebuyer willingness to sacrifice the glamour of active techniques to the simpler and possibly more reliable ones.[16]

**Passive Solar Home
Golden, Colo.**

Perhaps the biggest reason for the shift away from active systems — certainly in commercial buildings — is that they are most appropriate for buildings whose energy needs are mostly for heat. While heating typically accounts for about three-fifths of a home's total energy bill, it accounts for only about a third of a commercial building's energy budget. Michael Maybaum, former director of the Department of Energy's Passive and Hybrid Solar Division, believes "passive is far enough along" it can survive without government help.[17] According to the Solar Energy Research Institute, "the basic requirements for a successful commercial industry are in place: a well-developed commercial infrastructure and a market in which the new products can compete successfully." [18]

One important development that will probably be used widely in passive solar designs is heat mirror glazings. These glazings have become commercially available in several forms recently. They allow sunlight to enter a heated space but reduce heat loss in the daytime and — more important — at night, by reflecting infrared radiation back into the building. According to solar scientist Melvin Simmons at General Electric, scientists have

[15] Quoted in *The Wall Street Journal*, Feb. 24, 1982.
[16] See Christopher Flavin, "Energy and Architecture: The Solar and Conservation Potential," Worldwatch Institute, 1980, and "Trends in Architecture," *E.R.R.*, 1982 Vol. I, pp. 45-64.
[17] Quoted in *The Washington Post*, Sept. 5, 1981.
[18] SERI, "New and Renewable Energy in the United States of America," 1981, p. 15.

been working on heat mirror coatings for several years, particularly on cost, durability and transparency.

Focus on Solar Ponds as Cheap Fuel Source

For low-temperature industrial applications of solar energy, solar ponds may turn out to be the cheapest source, some experts believe. Scientists at Southern California Edison Co. and the California Institute of Technology's Jet Propulsion Laboratory (JPL) in Pasadena believe that such ponds can generate a steady supply of electricity at a cost comparable with natural gas or oil. Edison and JPL are working on a solar pond project at the Salton Sea, a 360-square-mile salt lake 140 miles southeast of Los Angeles. The project's director, Robert L. French, said a solar pond can generate electricity at 25 percent of the cost of any other solar technology. "If we are really serious about solar technology," he said, "this certainly is the best option we have today." [19]

Generally when bodies of water soak up heat from the sun the warm water rises to the surface and heat is released by evaporation. But in a pond infused with salt, a heavy layer of salty water stays at the bottom and traps the heat, initially absorbed by mud or some other material on the pond floor. The salty bottom layer of water inhibits the circulation that ordinarily carries the heat upward. The heated water at the bottom can then be drawn off. While this water will not usually reach the boiling point (212° F), it can become hot enough (180° F) to vaporize Freon 114 gas. The vapor can be used to drive electricity-generating turbines.

Edison's Salton Sea project is still in its design stages and will probably not be completed until 1984. The Department of Energy is putting up half the project's cost through this fiscal year, and the California Energy Commission and Edison are paying for the rest. "After successful completion of the experiment," project director French told Editorial Research Reports, "the technology, we hope, will be ready for the Southern California utility system to expand into a commercial plant."

French believes that under proper plant conditions an efficient pond could supply enough heat to generate power for two to four weeks without any sunny days. He predicts that ponds could supply 3 to 4 percent of the nation's total energy needs by the end of the century, twice the amount predicted for wind power.[20] There are drawbacks, however. Solar ponds have to be fairly close to what is being heated. They also require a lot of

[19] Quoted in "Solar Ponds: Saving Power for a Rainy Day," *Business Week*, April 20, 1981, p. 40.
[20] See "Wind and Water: Expanding Energy Technologies," *E.R.R.*, 1981 Vol. II, pp. 857-876.

land: to produce the 5,000 kilowatts of electricity needed for a community of 3,500 people, for example, a pond would have to cover almost half a square mile. Nearby salt sources are also desirable, since the ponds — generally 6 to 20 feet deep — have to be three times as salty as ocean water, which is 3.5 percent salt.

While there are only a few solar ponds in this country, Israel has several. For almost two years the Israelis have operated the world's first commercial pond, on the Dead Sea, and are now constructing a much larger plant they hope to have in operation by next year.

Largest Solar Power Plant to Open Soon

There have been unsuccessful attempts since the early 1900s to power steam engines with solar energy. Now, however, new concentrating techniques and storage materials are making high-temperature solar collection more practical for industrial and commercial uses. Collectors for high-temperature solar thermal systems consist mainly of devices that track and focus the sun's rays.

A "power tower," for example, consists of a series of flat mirrors, called heliostats, placed in such a way that they reflect the sun's rays to a fixed, central receiver, usually supported by a tower. SERI predicts that central receiver systems may be the cheapest way to produce large amounts of power if the technology proceeds at its present rate and that utilities may be an important future user of central receivers. One federal project in which the Southern California Edison utility company is involved is a central receiver plant in Barstow, Calif., called Solar One. Conceived in the early 1970s and expected to begin operation this year, Solar One will be the world's largest solar power plant.

The cost of Solar One is estimated to be about $123 million, of which $14 million is being provided by Southern California Edison, which will use the electricity from the plant and take it over when federal funding ends next year. Gerald Braun, director of the Department of Energy's Solar Thermal Energy Systems Division, told Editorial Research Reports that "getting the 10-megawatt pilot plant [at Barstow] operable remains the highest priority in our fiscal 1982 and projected 1983 program."

The first "total energy system" will also start operating in 1982, at a textile mill near Atlanta, Ga. This plant will simultaneously produce electricity and industrial process heat. Some experts predict that such cogeneration — where "waste" heat from one process, such as electricity generation, is used for another — will become an increasingly significant aspect of

Active Solar Homes, Denver, Colo.

high-temperature solar systems. It is further predicted the main use of solar thermal systems will be to generate steam to make electricity or to produce hydrogen-based fuels.

Progress in Photovoltaics — the Solar Cell

Solar advocates have described photovoltaics — a process that transforms the sun's rays directly into electricity — as "elegant" and clean. The technology involves no moving parts and is safe, reliable, efficient and simple to operate and maintain.[21] Of all the research programs in the proposed 1983 federal solar budget, photovoltaics will receive the largest chunk. Major electronic and oil companies are also investing heavily in photovoltaics.

The most serious problem with photovoltaics is expense: the process yields power 10 times more costly than that generated by oil. Although the price has fallen in recent years — from $50-$60 per peak watt of electricity in 1976 to $15-$20 per peak watt today[22] — the price would have to decline substantially to compete with electricity generated by conventional means. A kilowatt-hour of photovoltaic electricity costs about $1, while U.S. consumers pay only 2 to 10 cents per kilowatt-hour for conventional electricity. Paul Maycock, former director of the Department of Energy's photovoltaic program, believes that by 1986 the cost of producing solar cells will be only one-tenth what it is now, making photovoltaics economically feasible for private use.

Scientists are excited because technological advances are being made and the price of photovoltaic cells continues to fall while most other fuel prices rise. Silicon, which is extremely expensive, has been the main ingredient of solar cells, but a new, more cost-effective silicon configuration, "amorphous" silicon,

[21] The converting mechanism in photovoltaics, the photovoltaic or solar cell, has been used in America's space program for many years to provide energy in satellites. The cells consist mainly of two thin layers of material, one of them a semiconductor such as silicon and the other a metal such as aluminum or silver. A semiconductor can be treated so that, when light strikes it, electrons flow across the two layers — the so-called photovoltaic effect — and generate electrical current. This current is drawn off through wires to operate electric motors or other devices and furnish light.
[22] Peak power is that which is available when the sun is directly overhead.

was discovered about five years ago. While many scientists responded skeptically to the claim of its discoverer, Stanford Ovshinsky, that amorphous silicon is photovoltaics' ticket to economic viability, Atlantic-Richfield and Standard Oil of Ohio invested $82 million into Ovshinsky's research.

Last year the government awarded a $14 million photovoltaics grant to Texas Instruments to work on its proposed solution to the problem of what to do when the sun doesn't shine. The company's plan involves shooting molten semicrystalline silicon through a nozzle to make tiny beads. "Treated carefully with impurities and immersed in a weak acid solution," explained Burt Solomon in *Science 82* magazine, "these beads work like miniature solar cells, producing smidgens of electricity that are fed directly into the acid solution. This causes a chemical reaction that separates the solution's hydrogen from its other components. Later the chemical parts can be rejoined in a device known as a fuel cell, which turns chemical energy back into electricity" when convenient to the user.[23]

Photovoltaic Cells

Currently photovoltaic systems have proved practical mostly in sites that need little energy and are far from a power line. Most applications of photovoltaic systems are federally subsidized. But according to Paul Maycock, "if we seriously begin to adopt photovoltaics now, as much as 30 percent of the nation's electrical energy can come from this source by the year 2000." [24] The largest market will be small power users, including individual homeowners and small villages and communities around the world with modest electrical needs and no connection to a power grid. Utilities could be using solar cells by 1990, the Solar Energy Research Institute predicts.

Despite a marked decrease in federal allocations,[25] the photovoltaic industry is expanding. Sales in the United States went to $40 million in 1980, doubling the 1979 figure, and are expected to exceed $80 million this year. More than half of the U.S. production has been for export, and the industry expects the foreign-sales share to increase.

---

[23] Burt Solomon, "Will Solar Sell?" *Science 82*, April 1982, p. 74.
[24] Quoted in *Sun Times*, Sept. 1981, p. 7.
[25] Federal funding for photovoltaic research and development: $110 million in fiscal year 1979; $151 million in 1980; $132 million in 1981; $74 million in 1982; and $27 million requested in President Reagan's budget for 1983.

Although the U.S. photovoltaic industry is the world leader, it worries about the effects of reduced federal funding and expanding commitments by foreign governments to research and development. There is fear that Americans might wind up buying millions of efficient Japanese photovoltaic cells, at the expense of local manufacturers. Bob Johnson, an analyst with Strategies Unlimited, a Mountain Valley, Calif., market consulting firm that specializes in alternative energies, contends that federal cuts might leave small businesses unable to continue developing their technology and remain in the marketplace.

Solar Lobby's Center for Renewable Resources predicts that without government assistance, small businesses will have little opportunity in the U.S. photovoltaics market because of the oil industry's involvement. Four companies control 86 percent of the U.S. photovoltaics industry, the center reports, and of those four, three are wholly or partly owned by oil companies.[26] "A conflict of interests arises when photovoltaics can be substituted for other energy sources controlled by the oil companies," wrote Barrett and Lyndon Stambler in the January-February issue of *Sun Times,* the Solar Lobby's magazine.

The center admits that while it has found no evidence that the oil industry has directly suppressed photovoltaics research, the motivation exists for an oil company to do so and thereby reap benefits from its interests in oil or such other energy sources as coal or uranium. But the oil industry claims photovoltaic interests are its own. "We believe that eventually photovoltaics will be very big, and we want to be a very big player in this game," said Tom Guldman, manager of inventions development for Standard Oil.[27]

Outlook for Industry's Survival

PESSIMISTS point to other negative influences on the U.S. solar industry besides federal budget cuts and the administration's suggestion to repeal energy tax credits. The industry has also suffered from sustained high interest rates, a temporary oil "glut," and, until recently, poor earnings. The renewable energy industry is particulary susceptible to fluctuations in interest rates because many of its products — domestic hot water systems or wind turbines, for example — require a large

---

[26] Solarex, in which Amoco's holdings are estimated at between 25 and 40 percent; Arco Solar, a subsidiary of the Atlantic-Richfield Oil Co.; and Solar Power Corp., a subsidiary of Exxon. Pilkington Brothers, a British glass manufacturer, owns 80 percent of the fourth company, Solec International.
[27] Quoted by Burt Solomon, *op. cit.,* p. 71.

outlay of money. They have to be financed unless customers can pay cash. When interest rates are as high as they have been,[28] would-be buyers of solar systems often conclude that they cannot afford to buy what ultimately will save them money.

The biggest factor influencing the growth rate of the renewable energy industry may be the price of gasoline. Don Best, associate editor of *Solar Age,* writes that consumers often take their cues from the price of gasoline simply because it is the energy form most familiar to them. Dealers of solar-energy materials report that every time gasoline prices jump, their sales increase. And when gas prices ease, as they have in recent months, solar energy sales taper off.

While more than 300 U.S. companies manufacture solar equipment, production is highly concentrated in a few and the viability of small companies is tenuous. The nine largest manufacturers produce 94 percent of the low-temperature flat-plate collectors, for example. Many experts predict that this market concentration will increase as the less successful companies leave the field. Right now, according to Melvin K. Simmons of General Electric, most solar companies are consuming all of their investment capital "and will require both additional resources and a strong desire to stay in the market under present business conditions." Exxon, Olin Brass, Libbey-Owens-Ford and Standard Oil of California are among the large corporations that have dropped out of the solar business in the last year.

Rays of Hope Despite the Current Eclipse

Not everyone's outlook is grim. Martin E. Enowitz, president of Energy Investment Research Inc. in Greenwich, Conn., is one who predicts that solar energy will be the largest growth industry in the 1980s. He points out that studies and polls confirm that people want to see solar energy developed more than any other energy source. Moreover, solar companies have improved their customer service, solar energy is playing a larger role in the energy planning of countries around the world, and the technology is advancing rapidly. Enowitz also cites 1981 profits of several solar companies that had suffered big losses the previous year.[29] While some large firms are getting rid of solar subsidiaries, most are sticking by them.

The president of the Solar Energy Industries Association, R. Nicholas Loope, believes the industry could grow by 15 to 20 percent in 1982 if the recession is not too bad, and a San Francisco investment banking firm, Hambrecht and Quist, pre-

---

[28] The prime rate — the rate big commercial banks charge favored customers, usually large corporations — is currently about 16.5 percent. It had been above 20 percent in recent months.

[29] See Martin E. Enowitz, "1982: A Turning Point?" *Sun Times,* January-February 1982, p. 23.

dicts that "the alternative energy industry should be one of the fastest growing industries of the 1980s." The firm thinks that sales of renewable energy companies could climb to $66.5 billion in 1990. Total sales were about $7.8 billion in 1980, it estimates. In the end, according to the same appraisal, "the companies with capable managment and good products will evolve as the leaders and the other companies will either go out of business or will be acquired."

Industry leaders also hope to increase an already promising trade with other countries. In 1980, U.S. industry solar exports totaled $36 million, up 57 percent from the previous year, according to a Solar Energy Information Services report. The authors of the report estimated that in 1981 sales abroad would reach $65 million.[30]

Although President Carter set a national goal that 20 percent of this country's energy needs would be met by renewable sources by the year 2000, the Reagan administration projects that the contribution of renewable sources will grow from 2

---

[30] See Justin A. Bereny and Tom Jacobius, "Study of U.S. Solar Industry Exports, 1979-80," Solar Energy Information Services, 1981.

percent to only about 5 percent by then. While solar energy's future in America is clouded by revised projections and a change in national energy policy, the importance of solar energy's ultimate contributions is not in doubt. The basic questions are how soon the abundant energy from the sun can be harnessed efficiently enough to meet a substantial portion of the world's energy needs, and whether the United States will be in the forefront of that effort.

Selected Bibliography

Books

Brown, Lester R., *Building a Sustainable Society,* W. W. Norton and Worldwatch Institute, 1981.

Maycock, Paul D. and Edward N. Stirewalt, *Photovoltaics: Sunlight to Electricity in One Step,* Brick House Publishing (Andover, Mass.), 1981.

Solarex Corp., *Making and Using Electricity from the Sun,* Tab Books (Blue Ridge Summit, Pa.), 1979.

Solar Energy Research Institute, *A New Prosperity: Building a Sustainable Energy Future,* Brick House Publishing, 1981.

Articles

Congressional Quarterly Weekly Reports, selected issues.

"Removal of Tax Credits Would Devastate Industry," *SEIA News* (newsletter of the Solar Energy Industries Association), November 1981.

Simmons, Melvin K., "Solar Energy Technology — A Five-Year Update," *Annual Review of Energy,* 1981.

Solar Age, selected issues.

"Solar Ponds: Saving Power for a Rainy Day," *Business Week,* April 20, 1981.

Solomon, Burt, "Will Solar Sell?" *Science 82,* April 1982.

Sun Times (magazine of the Solar Lobby), selected issues.

Reports and Studies

Bereny, Justin A. and Tom Jacobius, "Study of U.S. Solar Industry Exports, 1979-80," Solar Energy Information Services, 1981.

Editorial Research Reports: "Solar Energy," 1976 Vol. II, p. 823; "Energy Policy: The New Administration," 1981 Vol. I, p. 57; "Trends in Architecture," 1982 Vol. I, p. 47.

Flavin, Christopher, "Energy and Architecture: The Solar and Conservation Potential," Worldwatch Institute, 1980.

"The MacNeil-Lehrer Report," PBS-TV, July 7, 1981.

Moore, J. Glen, "Solar Power" and "Solar Energy and the Reagan Administration," Congressional Research Service, Library of Congress, 1982.

Solar Energy Research Institute, "New and Renewable Energy in the United States of America," study conducted for U.S. Departments of Energy and State, 1981.

U.S. Department of Energy, "Solar Energy Update" and "Sunset Review" (3 vols.), 1982.

Illustrations on cover and p. 69 by George Rebh; on pp. 75, 83 by Staff Artist Robert Redding; photos by Solar Energy Research Institute

WOOD FUEL'S DEVELOPING MARKET

by

Marc Leepson

Oct. 16
1 9 8 1

Editor's Note: The Reagan administration's proposed fiscal 1983 budget contained no funds for wood energy research and development. It also would eliminate the $22 million Congress appropriated last year for the Solar and Conservation Bank, which among other things, was supposed to offer low interest loans for wood energy investments *(see p. 90)*. No loans have been made so far. Reagan wants to kill the program, saying that the bank would be a specialized federal subsidy program that would benefit only a "fortunate few."

WOOD FUEL'S DEVELOPING MARKET

THIS WINTER more American homes will be heated with wood than with electric heat provided by nuclear power.[1] It is ironic that in the last quarter of the 20th century an ancient heat source has become an important fuel for millions of Americans. A century ago nearly all American homes, businesses and industries were heated with wood. The use of wood fuel peaked in the 1880s, when Americans began switching to coal *(see p. 91)*. Later, oil and natural gas came to dominate the home heating market. By the early 1970s only about 1 percent of American homes used wood as the primary source of heat. But the decline in wood fuel use reversed after the 1973-74 Arab oil boycott.

As the cost of heating oil, natural gas and electricity soared, wood burning for heat began making a comeback. Today the United States is "on the crest of the wave of nations returning to wood," wrote Nigel Smith of Worldwatch Institute.[2] More than a million homes use this renewable energy source as their primary fuel and four million other residences use wood as an auxiliary heat source. Wood accounts for about 3 percent of the nation's residential fuel supply; in heavily forested parts of the country the percentage is much higher. Last year 20 percent of the homes in northern New England were heated solely with wood, and from one-third to one-half of the homes in the entire region burned wood for fuel. In Oregon, about 10 percent of the homes use only wood for heat, and about half have wood-burning stoves or fireplaces. In Georgia about one-fourth of the residences have wood stoves.[3]

The movement back to wood has been growing steadily since 1974, and analysts predict an even rosier future. A 1980 report by the congressional Office of Technology Assessment estimates that if fossil fuel prices continue to climb, as many as 10 million American homes might be using wood fuel exclusively or as an auxiliary heat source by 1985. The report also predicted that by

---

[1] Wood provided about 3 percent of all home heat in the United States in 1980. Fewer than 15 percent of all American homes are heated with electricity, and only a small percentage of that electricity is generated by nuclear power.

[2] Nigel Smith, "Wood: An Ancient Fuel with a New Future," Worldwatch Institute, January 1981, p. 16. Worldwatch Institute is a non-profit research organization in Washington, D.C.

[3] Figures provided by the Worldwatch Institute.

the year 2000 wood could meet as much as 20 percent of the nation's industrial and residential energy needs. "Although this represents more than a tripling of current use, it is nevertheless a minimum," the report said. "Much more energy could be obtained from this resource."[4]

Industry's Use of Wood Chips and Pellets

The movement back to wood also is taking place in industry. In fact, the forest products industry is the No. 1 user of wood fuel in the nation. Paper mills, sawmills and other processing plants use wood waste to fire combustion engines that produce electricity and steam. The industry as a whole is stepping up its use of wood fuel. In 1972 forest products plants received 40 percent of their energy from wood. Today that figure is approaching 50 percent, and it appears the trend will continue. The Office of Technology Assessment estimates that the industry could furnish 100 percent of its own energy needs by the year 2000.

A number of other industries use wood chips or pellets in boilers that produce electricity and heat. Matchbox-sized wood chips are made by machines that shred trees and convey the chips to delivery vans. "Wood-chip machines are especially advantageous alongside logging operations since they can make a useful product from the debris left by the loggers," Nigel Smith wrote. "Wood chippers can also cull old, diseased or contorted trees from forests, thereby creating more space for commercial grade trunks."[5]

Wood chip boilers have been installed in commercial buildings, universities and even in power plants. Among the first to use this type of wood-burning technology was the Burlington (Vt.) Electric Department, which converted two coal-burning boilers into wood chip users. The publicly owned utility plans to have a 50-megawatt wood-fired generating plant in operation by 1983.

Another type of standardized wood fuel, the pellet, is smaller but has a much lower water content than the wood chip. Wood pellets are made from pulverized paper and sawmill waste that is dried and compressed under high heat and pressure. One advantage of wood pellets compared to wood chips is that the former, when burned, give off only infinitesimal amounts of atmospheric pollutants and creosote — the flammable substance that adheres to the insides of stoves, boilers, pipes and chimneys.

---

[4] The predictions were based on forecasts that U.S. wood energy use could reach 2 Quads by 1985 and 11 Quads by 2000. A Quad is one quadrillion BTUs (British Thermal Units), the energy equivalent of about 168 million barrels of oil. See "Energy From Biological Processes," U.S. Office of Technology Assessment, July 1980, pp. 145-146.

[5] Smith, *op. cit.*, p. 24.

U.S. Energy Consumption*

Year	Wood	Coal	Petroleum	Natural gas	Hydro-electric power
1870	2.9	1.0	—	—	—
1880	2.9	2.0	0.1	—	—
1890	2.5	4.1	0.2	0.3	—
1900	2.0	6.8	0.2	0.3	0.3
1910	1.9	12.7	1.0	0.5	0.5
1920	1.6	15.5	2.6	0.8	0.8
1930	1.5	13.6	5.4	2.0	0.8
1940	1.4	12.5	7.5	2.7	0.9
1950	1.2	12.9	13.5	6.2	1.4
1960	0.3	10.1	20.1	12.7	1.7
1970	N.A.	12.7	29.5	21.8	2.7
1980	1.5	15.7	34.2	20.4	3.1

*In quadrillion BTUs (British Thermal Units)
Sources: David A. Tillman, *Wood as an Energy Resource* (1978); U.S. Department of Energy

The basic wood pelletizing process was developed more than two decades ago. Today's most widely used process, Woodex, was patented in 1977 by Rudolf Gunnerman, a German immigrant living in the United States. Woodex systems have been franchised throughout the world. Shell Oil Ltd. holds a franchise in Canada, and there are licencees in Brazil, Finland and Turkey. Woodex of New England, based in Boston, plans to have about 40 processing plants operating in the northeast United States by 1985. The plants will sell pellets mainly to business and industry, but there are plans to provide the fuel directly to residential users.

Until a wide-scale system of manufacture and distribution of wood pellets is implemented, the use of wood as fuel will be limited. "The pelletizing process ... is the key to wood's competitiveness as an industrial boiler fuel," wrote Harvard University business Professor Modesto A. Maidique. "Only if wood is pelletized can it be used as a substitute in a somewhat modified coal-firing system, and pelletizing also simplifies transportation and storage."[6]

Legislation to Encourage Wood Fuel Use

In the immediate future the nation's wood-burning effort will have to make do without a strong commitment from the Reagan administration. President Carter was an outspoken supporter of wood burning. He had a number of wood-burning stoves installed in the White House and the presidential retreat at Camp David, Md., and set up several governmental programs to pro-

[6] Writing in *Energy Future* (1979), edited by Robert Stobaugh and Daniel Yergin, p. 201.

mote the use of wood fuel. President Reagan has been widely photographed chopping wood for use in his Santa Barbara, Calif., ranch. But it appears that the president is applying his hands-off governmental policy to the wood fuel issue.

Last year Congress enacted legislation, the Wood Energy Utilization Act, that authorized recovery of wood residues in national forests for use as fuel or wood products. The bill was signed into law Dec. 19, 1980 — too late for funds to be appropriated. The Reagan administration chose not to fund the program this year. The administration also wanted to eliminate funds for the Solar and Conservation Bank, an institution Congress set up in 1980 that, among other things, offers low interest loans for wood energy investments. President Carter had asked for $125 million in funds for the bank for the fiscal year that began Oct. 1. The Reagan budget allocated no funds for the bank, but Congress authorized $25 million.

Pot Belly Stove

The biggest blow to supporters of expanded wood fuel use was the administration's withdrawal of support for a proposal that would have given a 15 percent tax credit to homeowners who purchased wood stoves. The tax credit plan was included in the tax-cut legislation, drawn up by the administration, that was passed by the House July 29. But the Senate version of the bill did not contain the wood-stove tax-credit measure. When a reconciliation conference of House and Senate members agreed on a final tax-cut bill Aug. 1, the wood stove credit was left out.

According to the Wood Heating Alliance, a Washington, D.C., trade association that lobbied heavily for the measure, the administration included the tax-credit provision in the House bill to try to get support from Republicans from the Northeast for the overall tax-cut package. The alliance maintains that the administration "had no apparent philosophical commitment to the wood stove credit," and it became "a disposable item" when the "Northeastern congressmen who voted with the president in the House apparently failed to use their leverage to obtain a firm commitment that the wood stove credit would be retained in the conference."[7]

Some observers predict that the Department of Energy, which under the Carter administration ran programs promoting wood as an alternative fuel, will now abandon the programs. "Under Jimmy Carter [promoting wood fuel] was better than

[7] Wood Heating Alliance, *WHA Report,* August 1981, p. 3.

sliced bread," said Bill Bulpitt, chief of wood energy systems for the Georgia Institute of Technology. "Under Reagan it isn't worth anything."[8] According to Danny C. Lim, program manager of the energy conversion equipment branch of the Department of Energy, Bulpitt's negative assessment may be premature. "We still are promoting the use of wood energy," Lim said in an interview. "Our programs are continuing in many of the facets that were developed under the Carter administration. Some of these are not going to be continued, but research and development activities will go forth in terms of new energy fuels from wood resources, in terms of upgrading the efficiency of equipment, promoting safety and promoting clean burning to reduce emissions from wood-burning equipment."

Fall and Rise of Wood Fuel

ANTHROPOLOGISTS believe that humans were burning wood for fuel at least 500,000 years ago. About 150,000 years ago humans began using fires for cooking as well as for heating their caves and warding off wild animals. Wood was the primary fuel for all the early civilizations — water power was a distant second. The Greeks developed wood-fueled silver mining and smelting operations. The Cypriots burned wood to manufacture bronze weapons. Wood continued to be the world's primary fuel source through the Middle Ages.

But by the 1500s some countries in Europe began running out of wood. Wood shortages became particularly acute in England in the 17th century. That nation's search for alternative fuels was one of the factors that led to the widespread introduction of coal. English iron ore producers began substituting coal for charcoal in 1709. By the end of the 18th century most of that nation's industries had switched to coal. English brick manufacturers found that coal was a useful fuel for their ovens since bricks fired in coal-burning ovens were more fire resistant. This allowed people to burn coal on their hearths without jeopardizing their houses, and coal became the primary residential fuel.

Coal did not come into widespread industrial use in the United States until the late 19th century.[9] Before that American industry was dominated by small, widely scattered factories that used wood, water and wind power. In the first half of the 19th century wood accounted for about 90 percent of the fuel used in the United States for home heating and cooking.

---

[8] Quoted in *The Christian Science Monitor*, Sept. 4, 1981.
[9] See "America's Coal Economy," *E.R.R.*, 1978 Vol. I, pp. 281-300.

Industrial Fuel Consumption, 1850-1870

	1850	1860	1870
Coal	19.4%	30.5%	57.7%
Water	25.0	22.0	20.0
Wind	38.8	35.6	12.9
Wood	16.7	11.9	9.4

Source: Sam H. Schurr and Bruce C. Netschert, *Energy in the American Economy* (1960).

Two events following the Civil War contributed to declining industrial and residential use of wood: the rapid industrialization of the country and the expanding population. From 1850 to 1870, David A. Tillman wrote, ". . . fuel wood utilization peaked, coal production and consumption increased dramatically, commercial petroleum production began in Titusville, Pa., and total energy usage almost doubled. . . . While all fuels were used in increasing amounts during this period, wood lost significant ground on a proportional basis. This trend was particularly pronounced in the industrial arena, where coal assumed the lead over all renewable resources."[10] By 1900, wood met only 25 percent of U.S. energy needs.

Influence of the 1973-74 Arab Oil Embargo

Wood use steadily declined during the first seven decades of this century, reaching a low point of just under 2 percent of fuel consumed in 1972. The following year marked a significant turning point in this nation's energy use. "Our rediscovery of wood, like so many of our new ways of looking at energy, was accelerated by the Arab oil embargo of 1973," wrote University of Virginia physicist James S. Trefil. "Individuals began dusting off grandpa's wood stove or went out and bought a modern version to heat their homes."[11]

Sales figures from the wood stove industry reflected the heightened interest in wood heat that followed the disruption of imported oil in the winter of 1973-74. "Stove sales in this country were fairly consistent up to the first Arab oil embargo of '73-'74," said Andrew Shapiro, president of Wood Energy Research Corp., a Camden, Maine, consulting firm. "Statistics show that between 150,000 and 200,000 units were being sold a year prior to 1973. Then the embargo hit, and stove sales started to take off. From 1974 through 1978 there was a geometric increase in the number of stove sales on an annual basis." In 1979, a record 1.5 million stoves were sold in this country. Although sales dropped off to about 800,000 units last year — analysts predict a similar total for 1981 — the new interest in

---

[10] David A. Tillman, *Wood as an Energy Resource* (1978), pp. 9-10.
[11] Writing in *Smithsonian*, October 1978, p. 55.

wood fuel revitalized the stove manufacturing industry.

In 1973 there were only about 12 stove manufacturers in the United States. Today there are about 75, located primarily in the Northeast. The story of Vermont Castings, Inc., in Randolph, Vt., provides an example of the rapid growth of the industry. The company began selling wood stoves in 1975 when founder Duncan Syme designed and developed three models based on European designs. An immediate success, Vermont Castings today has about 420 employees, a new $5 million foundry and a line of coal-burning stoves to go with its three wood models. The company has doubled its sales annually since 1975. Last year Vermont Castings sold 50,000 stoves — nearly all by mail — for some $20 million. The models cost from $400-$800. The privately held corporation has become the biggest employer in rural Orange County, Vt. As far as profits are concerned, the business "evidently turns a tidy profit," *The Wall Street Journal* commented, "but won't say how much."[12]

Fireplaces vs. Stoves; Hardwood vs. Soft

As many Americans have discovered in recent years, the old-fashioned fireplace is extremely inefficient when it comes to providing heat. Fireplaces are pleasing to look at, but about 90 percent of the heat given off by the burning logs goes through the chimney. Moreover, the draft created by the rush of heat up the chimney carries warmed air from the house out into the cold. There are ways to make fireplaces more efficient. These include installing glass doors, improving the grate and investing in a self-contained fireplace insert.[13] At best these devices can improve a fireplace's efficiency to the point where 50 percent of the heat given off stays in the house.

A much more efficient way to heat with wood is with a free-standing stove. Most wood-burning stoves radiate 60-70 percent of their heat into the house. The most advanced models have a 90 percent efficiency rate. The best models cost more than $600; others are available for under $200. Experts advise that for safety's sake wood-burning stoves be hooked up by an experienced installer.

A major expense for home wood burners is the wood itself. Firewood sells for as much as $175 a cord in some cities. In rural or heavily forested areas a cord costs about $75. A cord, the standard unit of wood measurement, contains 128 cubic feet of wood, measured in a pile standing four feet high, four feet wide and eight feet long. A cord of well-seasoned hardwood can

---

[12] *The Wall Street Journal,* Sept. 9, 1981.
[13] Fireplace inserts essentially are wood stoves that fit inside fireplaces. Inserts heat cold air drawn in from a house, and then blow the warmed air back into a room. Nearly all fireplace inserts contain electric blowers to circulate the warmed air. Inserts cost from $500 to $1,000.

provide the energy equivalent of about 150 gallons of heating oil. It takes from three to eight cords of wood to heat a home during a typical winter.

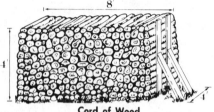

Cord of Wood

There are two basic types of firewood — hardwoods, cut from angiospermous trees such as oak and maple, and softwoods, cut from coniferous trees such as pine. Hardwoods burn longer, emit more heat than softwoods, and give off fewer pollutants. It is best to burn seasoned hardwood — wood that has been cut for at least three months. Seasoned wood has less moisture than fresh-cut (green) wood, and therefore gives off more heat because less energy is lost in the burning process.

Proper operation of a wood stove is essential to get it to burn efficiently. The most important factor is controlling the amount of air fed into the fire through the damper or air inlet. In essence, a small amount of air allows a fire to burn slowly, and larger infusions of air make the fire burn quickly. A slowly burning fire provides much less heat than a faster burning fire, but a rapidly burning fire can create a strong draft that pushes much of the heat out through the chimney. It takes time, patience and some expertise to learn how to operate a stove to its peak efficiency. As a U.S. Department of Energy publication advised: "All we can tell you is that the best position is somewhere in-between these two, and a little experimentation will help you discover where that position is for your particular stove."[14]

Growing Interest in Central Heating Units

Some industry experts believe that the future growth of the wood stove industry lies primarily with central heating units. "I have come to the conclusion that the stove end of the industry has peaked," Andrew Shapiro told Editorial Research Reports, "and I feel that the interest and growth in the area of solid fuels ... is in central systems." Shapiro sees a natural progression from oil burning to wood stoves and finally to central heating with wood. "During the 1975-79 period people had to reacquaint themselves with wood," he said. "There was up to a two generation gap of people who had been raised on the oil thermostat. Then people had stoves put in. They realized that you can heat and save money by using solid fuels."

The next step, Shapiro said, is getting rid of the stoves and installing a wood-burning furnace that ties in to existing hot

[14] U.S. Department of Energy, "Heating With Wood," May 1980, p. 7.

Heat Value of Wood

High (24-31 BTU* per cord)	Medium (20-24 BTU per cord)	Low (16-20 BTU per cord)
Live oak	Holly	Black Spruce
Shagbark hickory	Pond pine	Hemlock
Black locust	Nut pine	Catalpa
Dogwood	Loblolly pine	Red sider
Slash pine	Tamarack	Tulip poplar
Hop hornbean	Shortleaf pine	Red fir
Persimmon	Western larch	Sitka spruce
Shadbush	Juniper	Black willow
Apple	Paper birch	Large-tooth aspen
White oak	Redmaple	Butternut
Honey locust	Cherry	Ponderosa pine
Black birch	American elm	Noble fir
Yew	Black gum	Redwood
Blue beech	Sycamore	Quaking aspen
Red oak	Gray birch	Sugar pine
Rock elm	Douglas fir	White pine
Sugar maple	Pitch pine	Balsam fir
American beech	Sassafras	Cottonwood
Yellow birch	Magnolia	Basswood
Longleaf pine	Red cedar	Western red cedar
White ash	Norway pine	Balsam poplar
Oregon ash	Bald cypress	White spruce
Black walnut	Chestnut	

*British Thermal Units
Source: U.S. Department of Energy

water or hot air heating systems. A central heating unit in the basement eliminates the need to bring wood into the living areas of the house and to carry out ashes. In addition, central systems have large loading chambers that hold more wood than free-standing stoves. This means that central systems burn longer and require less frequent refills. Central systems also provide even heat distribution throughout a house, burn wood more thoroughly and emit fewer pollutants. Central systems cost from $1,000 to as much as $2,500.

Increasing Number of Fires and Injuries

As the number of wood-burning stoves and central heating units installed in American homes increases, so does the number of injuries and fires related to their use. According to the U.S. Consumer Product Safety Commission, the number of hospital emergency room burn injuries caused by wood stoves and fireplaces increased from about 600 in 1974 to some 4,600 in 1979, the last year for which complete statistics are available. The number of injuries doubled from 1978 to 1979. Experts estimate that 8,000 to 9,000 fires a year are caused by wood or coal stoves, resulting in about 100 deaths. More than 30,000 additional fires a year are thought to be caused by faulty chimneys, flues, or chimney connectors.[15]

The Insurance Information Institute in New York offers the following safety suggestions: (1) keep proper clearance between a wood stove and any combustible material such as walls, ceilings, furniture (especially wood bookshelves) and magazines; (2) check local building codes before installing wood stoves; (3) consult a detailed safety manual to be sure the stovepipe and flue meet safety requirements; (4) make sure the stove is made of a sturdy material such as cast iron or steel; (5) look for an Underwriters' Laboratories seal or other recognized testing laboratory approval before buying a wood stove; (6) check used stoves carefully for cracks or other defects in the legs, hinges, grates and draft louvers; (7) never use gasoline, kerosene or any other flammable liquid to start a fire and never use the stove to burn trash; (8) do not let a wood fire burn unattended or overnight; and (9) do not light the year's first fire without cleaning the smokepipe, elbow joints, flues and chimney.[16]

Environmental Consequences

WOOD FUEL USE on a massive scale has two basic environmental consequences. First, if tree cutting is not properly managed, it adds to deforestation — the loss of forest land. Second, improper burning of wood emits toxic particulates that add to air pollution. Air pollution is a problem primarily in the urban areas of developed countries. Deforestation and its consequences are global problems that are most acute in developing countries in Latin America, Asia and Africa. About 80 percent of all wood used in developing nations is burned for fuel, and experts say that the rate of tree planting must be increased fivefold in these areas to avoid severe wood shortages by the year 2000.

Deforestation has reached "alarming proportions in the Third World," wrote Nigel Smith of Worldwatch Institute. "Tropical forests are disappearing at the rate of 10 to 25 million hectares [about 24.7 to 61.8 million acres] a year. At the upper range of this estimate, which is an entirely possible figure, an area the size of West Germany is being annually stripped of its tree

---

[15] See Beatrice Harwood and Paul Kluge, "Hazards Associated with the Use of Wood or Coal-Burning Stoves or Free-Standing Fireplaces," U.S. Consumer Product Safety Commission, February 1980.
[16] Press release issued Nov. 3, 1980.

Estimated Changes in World's Forested Areas

Region	1978	2000	Change
	(million hectares)*		*(percent)*
U.S.S.R.	785	775	−1
Europe	470	464	−1
Japan, Australia and New Zealand	69	68	−1
Subtotal (Developed Countries)	1,464	1,457	0
Latin America	550	329	−40
Africa	188	150	−20
Asia and Pacific	361	181	−50
Subtotal (Developing Countries)	1,099	660	−40
World total	2,563	2,117	−17

*One hectare equals approximately 2.47 acres.

Source: Council on Environmental Quality, *Global 2000 Report* (1980)

cover."[17] The Council on Environmental Quality estimated that some areas of Asia and the Pacific islands would lose 50 percent of their forests between 1978 and 2000 *(see box, above)*.

The situation is not so dire in the United States, which along with Canada, northern Europe and the Soviet Union, has the heaviest concentration of coniferous and mixed hardwood forests in the world. Forests occupy 740 million acres in the 50 states, nearly one-third of the nation's land area. About 487 million acres, or some 66 percent of the forested lands, are classified as "commercial forests" — suitable for "continued production of large volumes of high quality timber" — by the U.S. Forest Service.

About 200-250 million of those acres are currently being used commercially, mostly for lumber. Of the 487 million acres classified as "commercial forests," the federal government controls about 107 million. State and other public forests occupy an additional 30 million acres, while the timber industry controls some 68 million. By far the largest amount of commercial forest — some 283 million acres — is in the hands of small, private land owners.

One of the most densely forested regions of the United States is New England. About 80 percent of the land area of Maine, Vermont, New Hampshire, Massachusetts, Connecticut and Rhode Island — some 32.5 million acres — makes up what is known as the Yankee Forest. Maine and New Hampshire have

[17] Smith, *op. cit.*, p. 7.

the heaviest concentrations of forested land in the nation — Maine is 90 percent forested, New Hampshire, 84 percent. Some 85 percent of the Yankee Forest is owned by about 500,000 individuals; the average plot size is 10 acres.

Encouragement of Proper Forest Management

Environmentalists have expressed concern that a widespread effort to tap U.S. forest lands through the use of whole-tree harvesting — also known as clear-cutting — could cause severe ecological damage. "A wholesale onslaught on the nation's forests to meet the short-term energy need would be an environmental disaster," said Richard Pardo of the American Forestry Association.[18] Clear-cutting can result in soil erosion that can turn forests into infertile wastelands. John G. Mitchell of *Audubon* magazine characterized widespread clear-cutting as "macro-destruction" of the forests. Mitchell expressed fears about "harvest technologies capable not only of large-scale clear-cuts, not only of slurping up whole trees and chipping them into vans, but of vacuuming *all* of a forest's biomass into the ever-expanding energy bag; of public utilities that know everything of generating kilowatts but absolutely nothing of regenerating trees. . . ."[19]

This environmental nightmare does not have to come to pass if proper forest management techniques are employed. One acre of woodland can provide a half cord of wood per year in perpetuity if proper harvesting techniques are employed. Among the ways to harvest wood in an environmentally sound manner are: (1) coordinating clearing and replanting strategies; (2) converting areas from hardwood to faster-growing softwood trees; (3) thinning regularly and (4) implementing disease, fire and insect control techniques.

Problems of Wood Smoke and Air Pollution

The second potential environmental drawback of wood fuels is the effect of wood smoke on air quality and human health. Burning wood gives off a number of potentially dangerous chemicals, including benzopyrene, sulphur dioxide and nitrogen dioxide. In addition, burning wood's particulate emissions — ash and soot — are proportionately greater than those emitted by oil and most types of coal, and can cause unsightly haze. Vail, Colo., recently enacted an ordinance limiting each newly built house to one wood stove because of severe haze caused by wood ash and soot. Wood fires are banned in London and in a number of South Korean cities for the same reason.

Carbon dioxide is emitted when wood, oil, natural gas, coal or

---

[18] Quoted in *The Progressive*, February 1981, p. 43.
[19] John G. Mitchell, "Whither the Yankee Forest," *Aububon*, March 1981, p. 79.

any other hydrocarbon is burned. But a higher percentage of carbon dioxide is given off by wood burning than by any other fuel. Scientists are concerned about recent increases in atmospheric carbon dioxide caused by combustion and deforestation — what is known as "the greenhouse effect." When excess carbon dioxide enters the atmosphere, physicist James S. Trefil explained, it "reflects heat back to Earth that would normally be radiated out into space, much as the glass roof of greenhouse traps heat."[20] The fear is that a long-term buildup of carbon dixoide in the Earth's atmosphere will cause severe damage in world weather patterns.

There are steps that those who burn wood can take to reduce air pollutants significantly. Moreover, all the steps that reduce pollution also increase the heating efficiency of wood stoves. The most effective method is to use a wood stove equipped with a catalytic converter, a device that can cut pollution by up to 66 percent. Catalytic converters for wood stoves are relatively new inventions that work in nearly the same manner as the automobile anti-pollution devices of the same name. But as energy analyst Kennedy P. Maize pointed out, "unlike the highway version, the converter for a stove actually improves operating efficiency."[21]

The catalytic converter in a wood stove releases heat into the house, and at the same time consumes pollutants that would otherwise have escaped through the chimney. Stoves equipped with these devices cost from $200-$300 more than wood stoves without catalytic converters. Replacements — converters wear out in about four years — cost $80-$90.

Another way to reduce air pollution from burning wood is to leave a stove's combustion air intakes open. This causes an almost smokeless fire, and all chemical emissions are burned completely. Other pollution control steps include burning only hardwoods that have been seasoned for at least six months, using medium-sized pieces of wood, not packing the stove too tightly, refueling often and never burning trash. "New information and technology — including pollution control devices — can keep wood from getting as bad an ecological reputation as oil, nuclear power, coal and other conventional fuels," wrote environmental writer Robert Deis. "By taking all possible precautions in the meantime, wood burners will be doing a great deal to make wood a cleaner fuel."[22] They will also be contributing, albeit in a small way, to the goal of reducing consumption of increasingly expensive fossil fuels.

[20] Trefil, *op. cit.*, p. 60.
[21] Writing in *New Shelter*, September 1981, p. 35.
[22] Writing in *Environmental Action*, December 1980, p. 7.

Selected Bibliography

Books

Clawson, Marion, *Forests for Whom and for What?* Johns Hopkins University Press, 1975.

Earl, Derek E., *Forest Energy and Economic Development,* Clarendon Press, 1975.

Fisher, John C., *Energy Crises in Perspective,* Wiley, 1974.

Gay, Larry, *The Complete Book of Heating with Wood,* Garden Way, 1974.

Stobaugh, Robert and Daniel Yergin, eds., *Energy Future,* Random House, 1979.

Tillman, David A., *Wood as an Energy Resource,* Academic Press, 1978.

Articles

Deis, Robert, "Where There's Wood There's Smoke," *Environmental Action,* December 1980.

Harris, Michael, "Farewell, Forests," *The Progressive,* February 1981.

High, Colin, "New England Returns to Wood," *Natural History,* February 1980.

Maize, Kennedy P., "Converting to a Wood Stove," *New Shelter,* September 1981.

Meyer, Amos J. II, "Fueling the 21st Century," *The Chase Economic Observer,* May-June 1981.

Mitchell, John G., "Whither the Yankee Forest," *Audubon,* March 1981.

"Nose-Thumbing at OPEC — With Wood Stoves," *U.S. News & World Report,* Jan. 21, 1980.

Trefil, James S., "Wood Stoves Glow Warmly Again in Millions of Homes," *Smithsonian,* October 1978.

Reports and Studies

Editorial Research Reports: "Forest Policy," 1975 Vol. II, p. 865; "New Energy Sources," 1973 Vol. I, p. 185.

Harwood, Beatrice and Paul Kluge, "Hazards Associated with the Use of Wood or Coal-Burning Stoves or Free-Standing Fireplaces," U.S. Consumer Product Safety Commission, February 1980.

Smith, Nigel, "Wood: An Ancient Fuel with a New Future," Worldwatch Institute, January 1981.

U.S. Department of Commerce, "Residential Energy Uses," 1979.

U.S. Department of Energy, "Heating With Wood," May 1980.

U.S. General Accounting Office, "The Nation's Unused Wood Offers Vast Potential Energy and Product Benefits," March 3, 1981.

U.S. Office of Technology Assessment, "Energy From Biological Processes," July 1980.

Wood Energy Research Corp., "1981 Woodfired Energy Systems Directory," 1981.

WIND AND WATER: EXPANDING ENERGY TECHNOLOGIES

by

Marc Leepson

Nov. 20
1 9 8 1

Editor's Note: President Reagan's proposed fiscal 1983 budget called for large cuts for all forms of solar energy research and development, including the wind energy program. The budget gave the wind energy program just $5.5 million, a stark contrast to the 1980 Wind Energy Systems Act which envisioned governmental expenditures on wind research to be $145 million in fiscal 1985 *(see p. 112)*. No funds were in the budget for hydropower research.

On Jan. 22, 1982, the U.S. Court of Appeals in Washington, D.C. struck down the regulations set up by the Public Utility Regulatory Policies Act of 1978 requiring public utilities to buy power produced by homeowners with windmills and small-scale hydroelectric producers *(see box p. 106)*. The government has not yet announced whether it will appeal.

WIND AND WATER:
EXPANDING ENERGY TECHNOLOGIES

THE UNITED STATES has been on a high-fossil-fuel diet for most of the 20th century. At the turn of the century coal was king. Then came oil and natural gas. Nuclear power was to be the fuel of the future.[1] But energy experts now predict that the nation's energy future will also include significant amounts of renewable fuel sources. For example, Virginia Electric and Power Co. (VEPCO), once a staunch advocate of nuclear power, is now considering the use of wind, water, solar power, wood, peat and municipal refuse to produce electricity.

VEPCO President William W. Berry announced at a Nov. 10 press conference that the utility was beginning an "unbiased, systematic search for the lowest cost option or combination of options for meeting future demand." First priority, he said, will go to reducing demand through conservation and "load management," or storing power in off hours for use in peak periods. To generate any additional power needed, Berry said, alternative fuels would be considered if VEPCO found them to be economically and environmentally acceptable.

Millions of homeowners and businesses throughout the United States already have begun to use alternative energy technologies. For example, wood now provides about 3 percent of the nation's residential fuel supply.[2] Interest in solar energy is growing steadily as the technology advances and the cost of hardware declines.[3] Wind power — an indirect form of solar energy since it is created by air heated by the sun's rays and cooled by their absence — may soon provide significant amounts of electricity in this nation and around the world.

The main advantage of wind power is that it constitutes a pollution-free, inexhaustible energy source. And, after the initial construction costs, it is almost cost-free. But there are problems. The wind does not blow steadily enough at all times and in all places, windmill construction is costly and energy storage is expensive. To minimize these problems, utilities can use clusters of large wind machines (see p. 107).

---

[1] Petroleum accounted for 44.9 percent of the energy consumed in the United States in 1980; natural gas, 26.9 percent; coal, 20.4 percent; and nuclear power, 3.5 percent, according to the U.S. Department of Energy.
[2] See "Wood Fuel's Developing Market," E.R.R., 1981 Vol. II, pp. 753-768.
[3] See "Solar Energy," E.R.R., 1976 Vol. II, pp. 823-842.

Wind energy "is a rapidly expanding field with far more immediate potential than most people realize . . . ," wrote Christopher Flavin of Worldwatch Institute. "[T]he wind is an economically attractive source of energy in many regions of the world."[4] Interest in wind energy is expanding primarily because of strides made in recent years by private and governmental research teams studying wind generation of electricity. Many of the improvements in windmill design represent a marriage of space-age technology and centuries-old techniques of harnessing power from the wind. "Windmills today are for the first time becoming really practical hardware," said Ned Coffin, board chairman of Enertech Corp., a Vermont windmill manufacturer.

No one is predicting that windmills will solve all of the world's energy problems. But the latest advances in design have prompted some optimistic forecasts. Energy analysts estimate that by the end of the century as much as 20 percent of the world's energy needs could be provided by the wind. Large-scale wind machines hooked into electric utility grids in the United States could supply about 2 percent of the nation's electricity by the year 2000. "In some windy areas like Hawaii, it could go higher than 10 percent," said Edgar Demeo, solar power manager of the industry-funded Electric Power Research Institute in Palo Alto, Calif.[5]

Development of Wind Power Technologies

The wind has been used as an energy source for centuries. Historians believe that the earliest wind machines probably were primitive devices used to grind grain in Persia around 200 B.C. There is also evidence of crude wind-driven machines in China 100 years later. The earliest known windmills were built during the ninth century in Seistan, an area spanning western Afghanistan's southwest border with Iran, where in summer the winds often reached 100 miles per hour. "Seistanis turned their local meteorological scourge into a benefit," wrote historian Jean-Louis Bourgeois. "They were the first people to channel the wind's wild energy into work." According to Bourgeois, "the sails of the initial Seistani mills almost certainly rotated horizontally like the millstone, thus bypassing the need for a gear to translate vertical into horizontal torque."[6]

Windmills were introduced to Europe late in the 12th century, probably by returning crusaders. The machines were first used to grind grain, saw wood, make paper and drain water.

---

[4] Christopher Flavin, "Wind Power: A Turning Point," Worldwatch Institute, July 1981, p. 5. Worldwatch Institute is a Washington, D.C., non-profit research organization specializing in global problems.
[5] Quoted in *The Wall Street Journal*, April 7, 1981.
[6] Writing in *Natural History*, November 1980, p. 70.

Wind and Water: Expanding Energy Technologies

Windmill use peaked in Europe in the 17th century, when there were about 10,000 windmills operating in England and 12,000 in Holland.[7] Extensive use of wind-powered ships and windmills enabled Holland to become the world's most industrialized nation in the 1600s.

Windmills were a primary source of mechanical power for American industries in the 19th century. In the 1850s and 1860s

wind energy was the No. 1 source of industrial fuel. But by 1870 wind began to be replaced by coal and water power.[8] Wind energy made a modest comeback beginning in 1890 following the development in Denmark of the first machine to use wind to produce electricity. By 1910, Denmark had constructed some 1,300 electricity-producing wind generators. The concept soon spread to the United States where hundreds of thousands of windmills were built to pump water and make electricity, primarily in rural areas of the Great Plains. But the coming of hydro-power and rural electrification made windmills obsolete. Windmills rapidly declined in popularity beginning in the 1930s. By the 1950s they had become picturesque reminders of a vanished age.

Experiments to harness wind power did continue in the 20th century, but on a very small scale. Before World War II German engineer Herman Honnef proposed building wind towers to generate electricity, and there was some research work carried out in Britain, Denmark, France, Germany, the Soviet Union and the United States. "These efforts were unfortunately sporadic since there was no sense of urgency to sustain them," Christopher Flavin wrote. "The prevailing view of future energy trends left little role for a seemingly antiquated energy technology such as wind machines."[9] That attitude changed markedly after the 1973-74 Arab oil boycott.

Growing Market for Small Wind Machines

One of the prime benefactors of the new research effort has been manufacturers of small wind machines (those producing less than 100 kilowatts of electricity) used in homes, farms,

---

[7] See Walter Minchinton, "Windpower," *History Today*, March 1980.
[8] See Sam H. Schurr and Bruce C. Netschert, *Energy in the American Economy* (1960).
[9] Flavin, *op. cit.*, p. 10.

How to Get Your Electric Meter to Run in Reverse

There are two basic types of home windmills: those that produce electricity to be stored in batteries for later use, and those that send electricity directly into a home's electric power panel. Owners of the latter kind are eligible for an extra bonus thanks to a law Congress passed three years ago to encourage the use of renewable energy sources. The Public Utility Regulatory Policies Act, part of President Carter's 1978 energy bill, among other things prohibits state utility commissions and other regulatory agencies from "discriminating" against solar, wind, water or other small-power systems.

This means that utilities must allow homeowners with wind, water or solar energy producing machines to tie them into the home power panel. This, in turn, means that when the wind is blowing steadily the electric meter slows down. During especially windy periods it is possible that the meter actually will run backwards. If this occurs, the electric company has to pay a "fair rate" for the excess electricity.

factories and small businesses. In 1973 there were only a handful of manufacturers. Today there are nearly 30 such companies in existence, selling about 2,000 windmills a year. One of the most successful is Enertech Corp. of Norwich, Vt., the nation's largest manufacturer of electricity-generating windmills. Enertech started marketing existing wind machines in 1975. Three years later the company began making its own models.

Enertech's wind machines were among the first to pump electricity directly into the electric power panel of a home rather than into storage batteries (see box, above). "We got rather disillusioned with the traditional battery-charging windmills that everyone was producing and marketing," said Board Chairman Ned Coffin. "We decided we'd make a windmill that would generate electricity synchronous with utility line voltages. It wouldn't need a battery. It wouldn't need an inverter. It's basically designed to slow down the electric meter." A home electric meter will slow down and then run backwards if a windmill is generating more electricity than is being used by the homeowner.

Enertech has sold about 750 windmills (called home wind-energy systems) in the last three years. The smallest model, which has a rotor blade 13 feet in diameter, is designed for a house using about 700 kilowatt hours of electricity a month — one that does not have air conditioning or electric heat. These 1.8 kilowatt machines provide 200-500 kilowatt hours of electricity a month if the wind is blowing steadily at an average speed of at least 12 miles per hour. A larger, 20-foot rotor model is designed for a house with a number of electric appliances but that does not rely heavily on air conditioning or electric heat. The cost of buying and installing the smaller model runs from $6,500 to $10,000. The larger machines cost from $12,000 to $20,000. Homeowners purchasing windmills qualify for a 40 percent tax credit under legislation Congress passed last year to encourage use of renewable energy sources *(see p. 113)*.

Karl W. Plitt of Clarksburg, Md., recently installed a smaller Enertech model on a 50-foot telephone pole behind his house in a rural subdivision — a site chosen in part for its windy conditions. After installing the wind machine and a companion solar hot-water heater, the Plitt house saved about half ($90) on the first two-month electricity bill. Walter F. Szymczweski built a four-blade, 30-foot high wind machine for about $7,000 at his Severn, Md., house. His electricity is stored in more than 200 batteries. "This country is in trouble, that's why I had to do this," said Szymczweski. "I was priced out of the heating market."[10]

Expanded Role for Utility Consumption

Although the home market for windmills is growing rapidly, energy experts say that the type of wind technology that will most benefit the nation will be large machines that feed electricity to utilities. With blades as long as 300 feet, towers as high as 200 feet, and at least 100 kilowatt generating capacity, these machines dwarf the home models. Some large windmills are capable of generating more than one megawatt (1,000 kilowatts) of electricity. Used in clusters called "wind farms," large wind machines could soon play an important role in the nation's energy picture. The American Wind Energy Association predicts that the use of large-scale wind machines by utilities "... may soon be the most common application of wind energy."[11]

Dozens of utilities around the nation are experimenting with wind power. The leaders in the field are in California, Hawaii and Washington state, but there is activity in nearly all parts of the country. For example, Public Service Co. of New Hampshire, that state's largest utility, has signed a contract with U.S. Windpower Inc. of Burlington, Mass., to install and test 20 windmills on Crotched Mountain, N.H. The electricity generated from that wind farm will be sold to Public Service. The Virginia Electric and Power Co. announced Nov. 10 that it will

---

[10] Plitt and Szymczweski were quoted in *The Washington Post*, Sept. 8, 1981.
[11] "Wind Energy: An Introduction," American Wind Energy Association, 1980, p. 7. The American Wind Energy Association in Washington, D.C., is a non-profit corporation established in 1974 to promote wind energy.

How the Wind Blows
in Selected U.S. Cities

City	Average Wind Velocity (in MPH)
Albuquerque, N.M.	8.8
Bismarck, N.D.	10.8
Buffalo, N.Y.	12.6
Chattanooga, Tenn.	6.3
Chicago, Ill.	10.3
Cleveland, Ohio	10.9
Denver, Colo.	9.2
Detroit, Mich.	10.0
Galveston, Texas	11.1
Jacksonville, Fla.	8.8
Knoxville, Tenn.	7.4
Louisville, Ky.	8.3
Miami, Fla.	9.0
Mobile, Ala.	9.5
Nashville, Tenn.	7.5
New York, N.Y. (Battery)	14.5
New York, N.Y. (Central Park)	9.5
Philadelphia, Pa.	9.6
Portland, Ore.	7.8
St. Louis, Mo.	9.5
San Francisco, Calif.	10.4
Spokane, Wash.	8.4
Toledo, Ohio	9.5
Washington, D.C.	9.4
Mt. Washington, N.H.	35.4

Source: Ed Trunk, "Free Power From the Wind," in *The Mother Earth News Handbook of Homemade Power* (1974), p. 143.

soon begin a three-year, multimillion-dollar study to see what role wind and other alternative energy sources will play in the 1990s in its service area *(see p. 103)*.

The Hawaiian Electric Co. has placed an order with windmill manufacturer Hamilton Standard to buy 20 four-megawatt machines for a wind farm on the island of Oahu. The project, which would be the largest of its kind, will cost some $350 million, and is expected to provide about 9 percent of the island's electric power by 1985 — a saving of about 600,000 barrels of oil a year. Installation of the windmills will be handled by Windfarms, Ltd. of San Francisco, a company that serves as a link between manufacturers such as Hamilton Standard and utilities.

Windfarms also has signed a preliminary agreement with California's Pacific Gas & Electric Co. and that state's Department of Water Resources to build a 146-windmill, 350-megawatt wind farm on a 5,000-acre tract in Solano County northeast of

San Francisco. The project is scheduled to be completed in 1989, and will provide electricity for about 150,000 homes. Windfarms also is looking into setting up large wind farms in Oregon and Washington.

Large windmills have been criticized for their stark, hulking appearance as well as for the noises given off by their blades. A 2-megawatt experimental machine installed by the Department of Energy near Boone, N.C., generated complaints from nearby residents about the constant "thumping" noises of the twin 100-foot-long steel blades as they turned. Large windmills sometimes cause electrical interference, resulting in interruptions of television signals in the immediate vicinity. On Block Island, R.I., the site of an experimental Department of Energy windmill, the residents made the government pay for a cable TV system so local residents could receive uninterrupted television programming.

Considering wind energy's benefits, the noise and interference problems are relatively minor. Researchers nevertheless are working on new designs that eliminate blade noise and television interference. Another solution to the noise and aesthetic problems is to locate giant windmills in out-of-the-way places, including offshore sites.

Offshore windmills are "much less obtrusive," said Glyn England, chairman of Britain's Central Electricity Generating Board. "Admittedly, it will be more difficult and costly to build and maintain the machines offshore than on land. There will be the extra expense of getting the power ashore. And they could be a hazard to fishing vessels. But wind speeds are higher there than over open land, although not quite as high as on the best hilltop sites."[12]

Directions of Wind Research

THE U.S. GOVERNMENT has been a generous supporter of wind energy research and development. At least four federal agencies have been involved in the field. The Department of the Interior, which manages federally owned land in the West where winds are often strong, has conducted feasibility studies on integrating wind power with existing hydroelectric systems in Wyoming. The Department of Agriculture has studied the use of wind turbines with irrigation systems on farms in Texas and Kansas. The National Aeronautics and Space Administration

[12] Quoted in *The Christian Science Monitor*, Oct. 29, 1981.

(NASA), because of its aeronautical design expertise, is the federal agency in charge of windmill machine design. The U.S. Department of Energy (DOE) provides the bulk of federal funding. Since the department was formed in 1977 it has spent some $240 million on wind energy research.

Government contracts led to the development of the most commonly used large windmill, the MOD-2, which is manufactured by Boeing. Three of these 200-foot tall machines, each generating up to 2.5 megawatts, have been installed at Goldendale, Wash., in the Columbia River Valley, where they are plugged into the Bonneville Power Administration's electricity grid. The MOD-2 has two steel rotor blades 300 feet in diameter, which sit atop a 200-foot tubular steel tower. When the wind hits 14 miles per hour, the machines go into motion, reaching their rated power of 2.5 megawatts at 27.5 miles per hour. Three machines, when operating at peak efficiency, provide electricity for about 2,500 homes.

Industry analysts say that these large machines will be widely used by utilities beginning next year. "We see a small amount of activity in 1981, but we think it will be 1982 before it really takes off," John E. Lowe, Boeing's director of wind energy programs, said last spring. "We have proved that the system will do everything we said it will. We have not proved the lifetime (designed to be 30 years), but we think we have enough data to believe we will achieve it."[13]

Other manufacturers of large windmills include Bendix Corp., WTG Energy Systems, Alcoa Co. and Bensic Wind Power Products Co. Hamilton Standard, which has built the first factory to manufacture blades for large machines, hopes to produce two per month by the end of the year. "That would be quite an achievement and could revolutionize the market," wrote Chris-

---

[13] Quoted by Richard G. O'Lone in *Aviation Week & Space Technology*, March 23, 1981, p. 43.

Federal Funds for Wind Energy Research and Development

Fiscal Year	Millions of Dollars
1973-74	$ 1.8
1975	7.9
1976	14.4
1977	27.6
1978	35.5
1979	59.6
1980	63.4
1981	54.0
1982	18.3*

*Administration request pending congressional action.
Source: U.S. Department of Energy

topher Flavin. "In less time than it takes to plan and build a conventional power plant, the first large wind farms should be in operation."

One way to measure the technological progress of the federally funded research effort is to compare the costs of producing wind energy. The MOD-2 machines now produce one kilowatt hour of electricity for less than 10 cents. With the experimental machines of the early 1970s, the cost for a kilowatt hour of electricity was about one dollar. "We've gone from experimental units for which our return on the investment was significant advances in the technology — if not in the cost of energy — to a machine that reflected a bit of both, the MOD-2," said Kurt Klunder, acting deputy director of DOE's Division of Wind Energy. "And we have a MOD-5 on the drawing board that is projected to reduce that to something like 3-6 cents per kilowatt hour, the difference being the actual site and the wind resource that's available. ... Those kinds of numbers are ones that can make wind machines attractive and even commercially competitive."

Effect of Administration's Budget Cuts

Last year Congress passed and President Carter signed into law a bill, the Wind Energy Systems Act, designed to boost the federal windmill research and development program. The act established (1) a small wind energy systems program to encourage the windmill industry through federal procurement of windmills for use by government agencies, (2) a large wind energy systems program that would use federal research, grants and loans to boost development of large wind farms and (3) a three-year, $10 million assessment program to determine where the ideal sites are in the United States to harness wind energy.[14]

[14] See Congressional Quarterly's *1980 Almanac*, p. 489.

111

The congressional conference committee that drafted the final bill estimated that the government would spend $891 million on wind energy from fiscal years 1981-8. The projections were as follows:

Fiscal year	Millions of dollars
1981	$100
1982	173
1983	145
1984	133
1985	125
1986	80
1987	60
1988	75
Total	**$891**

There is little doubt today that the government will spend only a fraction of that total, and that nearly all of the Wind Energy Systems Act's programs will be scrapped. Reagan administration budget cutters wasted little time in trimming federal funds for wind energy research this year. The administration asked for and received from Congress a rescission in the fiscal 1981 wind research appropriation from the original $100 million to $54 million. For fiscal year 1982, which began Oct. 1, the administration first asked Congress to appropriate $19.4 million for wind research, and then lowered the figure to $18.3 million. Originally the budget had projected $173 million for wind energy research for fiscal 1982. Congress has yet to take final action on the funding request.

The administration believes it is now up to private enterprise to put into practice the knowledge gained in the last six years of federally promoted wind energy research. Tom Gray, acting executive director of the American Wind Energy Association, agrees at least in part. Gray said in a recent interview that previous government-funded research had paid off with positive results, especially in the area of large-scale windmill technology. "Even without the government program, within a few years we'll begin to see large-scale turbines up and around in a number of areas of the country," Gray said. "The small machine market has been growing pretty rapidly and I think that will continue. . . . The original purpose of the Wind Energy Systems Act was not necessarily that it would result in something happening that wouldn't eventually have happened anyway because sooner or later the price of oil is going to rise enough so that other sources become competitive."

Others believe that it is too early to cut back on government funding for wind research. "The job isn't quite done," said

Edgar Demeo of the Electric Power Research Institute in California. "I wouldn't say the government has to spend hundreds of millions of dollars more, but I have doubts that the government should pull out entirely at this point."[15]

Kurt Klunder of the Department of Energy explained how the funding cutbacks will affect the agency's existing programs. "We will no longer undertake development of total machines," he said. "We're seeing enough activity in the private sector now to stay out of the way and not favor one developer over another. We will try to concentrate our efforts on the technology-based research in concert with the administration's philosophy. We had undertaken a number of projects in a field evaluation program to get small machines in every state to get the state energy offices involved in siting, zoning and so on. These kinds of field projects will not be conducted anymore. . . . We will also be backing off somewhat in the amount of machine testing we're doing, although we hope to continue to test the MOD-2s because it takes two years or more to know what we've got there."

Future of Income Tax Credit Incentives

While there is disagreement on the effects of the cutbacks for research and development, there is widespread concern about the possible repercussions of an administration suggestion — hinted at in the Sept. 24 presidential budget message to Congress — to eliminate the special homeowner's tax credit for installing renewable energy systems. A provision in the windfall profits tax measure passed by Congress March 27, 1980, increased the tax credit available for residential solar, wind and geothermal equipment to 40 percent of the first $10,000 in expenditures — a maximum of $4,000. The Energy Tax Act of 1978 had provided a credit for 30 percent of the first $2,000 and 20 percent of the next $8,000 with a maximum credit of $2,200.[16]

Manufacturers of small windmills designed for home use say the tax credit has been a crucial factor in their industry's recent success. One reason is that the least expensive models cost about $6,000, and many homeowners cannot afford to buy windmills without a tax credit. A change in the current tax credit law, which is scheduled to expire in 1985, would be a "severe blow" to the small windmill manufacturers, as well as to the future of wind power in this country, said Tom Gray of the American Wind Energy Association.

In addition to the 40 percent homeowner tax credit, a 15 percent business investment tax credit also affects the small windmill business. "There are currently some investment deals

[15] Quoted in *The Wall Street Journal*, April 7, 1981.
[16] See Congressional Quarterly's *1980 Almanac*, p. 475, and *1978 Almanac*, p. 641.

being put together to construct large numbers of machines,"
Gray said. "A major part of that is the tax incentive involved in
it." Capitol Hill observers say there is strong sentiment in
Congress from both Republicans and Democrats to retain the
tax credits and that it is unlikely the administration would be
able to get enough votes to do away with them. Just bringing up
the subject, though, has had an impact, Gray said. "It hurts just
to have them talking about it. It creates instability in the
investment climate."

Promise of Hydroelectricity

UNLIKE WIND POWER, water power already accounts for
a significant portion of the energy consumed in the United
States and around the world. Hydro-power stands behind petroleum, natural gas and coal — and ahead of nuclear power — on
the list of energy sources consumed in the United States.[17]
According to U.S. Department of Energy statistics, hydroelectric power last year generated 4.1 percent of the energy consumed in this country and 12 percent of the nation's electricity.
Water power produces an estimated one-fourth of the world's
electric power and about 5 percent of all the energy consumed.

The United States leads the world in the production of hydroelectric power, although other industrialized nations including
the Soviet Union, Canada, Japan, France, Norway, Switzerland
and Austria also generate large amounts of water power. Many
Third World countries also are heavily involved in tapping
falling water to generate power. According to United Nations
statistics, 44 percent of the electricity used in the developing
world comes from hydro-power. Ghana, Zambia, Mozambique,
Zaire and Sri Lanka obtain more than 90 percent of their
electricity from water power.

Some of the world's largest hydroelectric projects are located
in developing countries, including Zimbabwe's 1,300 megawatt
Kariba Dam, Egypt's 2,100 megawatt Aswan High Dam, Brazil's
4,000 megawatt Furnas Project and Venezuela's 2,900 megawatt
Guri Project. The Itiapu Dam now being constructed on the
Parana River between Brazil and Paraguay will generate 12,600
megawatts — the amount of power produced by a dozen nuclear
power plants.[18]

The biggest hydroelectric project in the world is the Grand

---

[17] See footnote 1, p. 103.
[18] See Daniel Deudney, "Rivers of Energy: The Hydropower Potential," Worldwatch
Institute, June 1981.

Coulee Dam, located on the Columbia River in Washington state. The giant dam, which took nine years to build (1933-42), is about a mile long and 550 feet high. Its nine 165,000 horsepower turbines generate just under 6,500 megawatts of power — about a third of all the hydroelectricity produced in the nation.

Some energy experts are predicting an even larger role for water power in the future. Lester R. Brown, president of Worldwatch Institute, has said that water power once fully developed "could triple, or perhaps even quadruple existing generating capacity. While a few countries such as Japan and Switzerland have little undeveloped hydroelectric potential remaining, most countries are far from this point.... Overall, this source of renewable energy promises to become a cornerstone of a renewable society."[19]

From Water Wheel to Electric Generation

Water power has been employed as an energy source since the first century B.C. when the Greeks used water-powered grist mills to grind grain. Water wheels were used throughout the world for centuries, primarily to drive machinery in flour mills and factories. A series of discoveries by French engineers in the 19th century led to the development of hydroelectric power. The French — in contrast to the British and Americans who turned to the development of coal — concentrated on developing water power as the successor to wood fuel. The main reason was that nation's abundant rivers.

The first important breakthrough in the field was the invention of the water turbine by Benoit Fourneyron in 1820. Although Fourneyron is credited with inventing the turbine, another French engineer, Claude Burdin, coined the term in an 1822 academic paper. Burdin derived the word from the Latin *turbo,* which means that which spins.[20] More than 50 years after the invention of the turbine the first central hydro-generating facility produced electricity in Appleton, Wis. This initial use of hydro-power came when water-fed turbines were hooked to a generator and gave off enough electricity to light 350 light bulbs.

During the first decades of the 19th century water power played a crucial role in the widespread industrialization of the United States, especially in New England. Lowell, Mass., is generally regarded as the birthplace of the American industrial revolution. Soon after the city was established in 1826 a number

---

[19] Lester R. Brown, *Building a Sustainable Society* (1981), pp. 218, 222.
[20] See Norman Smith, "The Origins of the Water Turbine," *Scientific American,* January 1980, pp. 138-148. The idea that water power is France's equivalent of coal is borne out by the French term for water power — *houille blanche* — which literally translates as "white coal."

of industrialists from nearby Boston chose Lowell as the site for a large textile industry. The main attraction was the rapidly flowing Merrimac River. A series of canals was built and Lowell's factories soon were deriving all their power from the flowing water.

"It's no accident that New England was the birthplace of the industrial revolution in this country," noted environmental writer Michael Harris, "it seems almost anywhere you go in the six-state region there's a rushing river, a steadily flowing stream or an unnamed brook. In the 19th century, good old fashioned Yankee ingenuity put them all to work. Thousands of the area's streams were used to power sawmills, textile mills, machine factories, and later, the first street lights and electrified homes."[21] The use of water to provide industrial power peaked in the mid-19th century when hydro-power made up about one-fourth of all industrial fuel. Water power use dropped steadily in the late 1800s as industries began to use coal and then petroleum as their energy sources.

Government-Backed Dam Projects Out West

Nationwide use of water power declined until the 1930s. Then during the Depression the federal government began financing the construction of large dams through the New Deal's Works Progress Administration (WPA). One of the first projects was the Grand Coulee Dam in Washington *(see p. 115)*. Construction began in 1933 after Congress granted WPA $450 million for the job. One benefit of the nine-year construction period was that 12,000 workers were able to find jobs at a time when there was massive unemployment. In all, more than two-dozen government-financed dams were built during the 1930s.

In 1937 Congress set up the Bonneville Power Administration (BPA) to send and sell power from the federally constructed dams in the Columbia River Basin. The system provides hydroelectric power to Washington, Oregon and the western parts of Idaho and Montana. BPA supplies about half the region's electric power at rates that until recently were among the lowest in the nation. But beginning in the early 1970s electricity consumption in the region started to increase at a 3 percent annual rate, and that caused allocation problems.[22]

---

[21] Writing in *Environmental Action*, June 1979, p. 25.
[22] In an effort to remedy the situation, a group of 23 small public utilities, the Washington Public Power Supply System, began constructing five nuclear power plants in 1973. Serious cost overruns forced cancellation of two of the plants. The original cost was slated to be $4.1 billion. The current estimate has risen to $23.8 billion, and the three remaining plants will not be completed until 1984. The financial troubles spilled over into the U.S. municipal bond market due to the system's inordinately heavy borrowing to meet construction costs. The project's latest setback came Nov. 4 when Washington voters approved a ballot proposition prohibiting the system from issuing revenue bonds for power plants without voter approval. *Newsweek* magazine (Nov. 16, 1981) called the situation the "biggest public-power mess in U.S. history."

Dam Consequences

There is good news and bad news involved in building giant hydroelectric dams. The good news is that the dams provide the means for turning flowing water into electricity. The bad news is that the dams represent a human intrusion into nature — an intrusion that drastically changes the ecological balance of river systems. Dams change the flow of rivers, turn valleys into reservoirs and affect vegetation, fish and wildlife. If an altered river system's ecosystem is not properly managed after a dam is built, there can be serious consequences, including erosion and flooding.

The Reagan administration, which has been harshly criticized for its disregard of environmental concerns, has taken action in this area. Secretary of the Interior James G. Watt announced Oct. 29 that the government was dropping plans to add two new generators to the Glen Canyon Dam power plant on the Colorado River. The new generators would have caused a wide fluctuation in the flow of the Colorado River and other environmental consequences. "At issue," Watt said, "is a key environmental concern — the integrity of the Colorado River downstream from the dam as it flows through Grand Canyon National Park."

In 1973 BPA stopped selling power to the seven utilities in the region owned by private investors. Three years later the power administration told its preferred customers — a group of 116 public utilities — that it would not be able to meet their additional electricity demands after July 1983. The agency also told its industrial customers that power contracts with them were not likely to be renewed. The investor-owned utilities, mostly in Oregon, Montana and Idaho, were thus forced to produce electricity from coal or nuclear power plants — power that is much more expensive than the hydroelectricity provided by BPA.

Last year Congress enacted legislation to try to remedy the problems caused by expanding electricity demands in the Pacific Northwest. The bill, which President Carter signed into law Dec. 5, 1980, established a four-state regional council to (1) allocate power to federal dams in the area, (2) promote energy conservation and (3) acquire new energy sources to meet the section's power needs. Congress asked the Bonneville Power Administration to come up with a comprehensive regional energy plan that would emphasize conservation first, renewable energy sources second, and finally, the acquisition of additional coal or nuclear power plants.[23]

According to Nicholas Dodge, chief of water management for the U.S. Army Corps of Engineers' Pacific Northwest Division,

---

[23] See Congressional Quarterly's *1980 Almanac*, p. 486.

U.S. Consumption of Hydropower*

Year	Quadrillion BTUs	Billion Kilowatt-Hours	Percent of Total Energy Consumed
1950	1.44	102.7	4.3
1951	1.45	106.6	4.0
1952	1.50	112.0	4.2
1953	1.44	111.6	3.9
1954	1.39	114.0	3.9
1955	1.41	120.3	3.6
1956	1.49	129.8	3.7
1957	1.56	137.0	3.8
1958	1.63	146.9	4.0
1959	1.59	144.7	3.7
1960	1.65	153.7	3.7
1961	1.68	157.5	3.8
1962	1.82	172.2	3.9
1963	1.77	169.1	3.6
1964	1.91	182.3	3.8
1965	2.06	196.8	3.9
1966	2.07	199.0	3.7
1967	2.34	224.6	4.0
1968	2.34	225.2	3.8
1969	2.66	254.5	4.1
1970	2.65	252.9	4.0
1971	2.86	273.1	4.2
1972	2.94	283.6	4.1
1973	3.01	289.7	4.0
1974	3.31	316.9	4.5
1975	3.22	309.3	4.6
1976	3.07	295.5	4.1
1977	2.51	241.0	3.3
1978	3.16	303.2	4.0
1979	3.17	303.4	4.0
1980**	3.13	299.5	4.1

Source: U.S. Department of Energy

*Electric utility and industrial generation of hydropower.
**Preliminary

BPA's hydroelectric system eventually will be used primarily as a "peak" supplier of electricity. The dams will generate electricity only during periods of great demand, and will serve mainly as a backup for the region's coal, nuclear and oil-powered power plants.[24]

Bright Future for Small Hydro-Power Dams

While the future of large-scale hydro-power plants in this country is somewhat clouded, energy experts believe that many

---

[24] See *Seattle Times*, Oct. 11, 1981. The U.S. Army Corps of Engineers, charged with the responsibility for flood control and navigation in the United States, controls the distribution of water from the Columbia River to the federally owned hydroelectric plants in the system.

long-abandoned small dams could be resurrected without too much expense. Thousands of dams have been abandoned in the last 50 years, mostly in the Northeast and Midwest. A report issued in Janaury 1981 by the New England River Basin Commission found there were about 1,750 unused small dams in the region, and that about half of them could be renovated easily to produce power today.

"New England started on these things," said Louis Klotz, a civil engineering instructor at the University of New Hampshire. "These were small plants designed to provide for local needs. Now we have centralized power supplies and distribution lines, and right in our backyards is all this abandoned power. Doesn't it make sense to put it to use to create local jobs so people won't have to use energy to commute back and forth to work? We could use this power to run small and medium sized industries and market the power, in effect, through the products we could be making locally."[25]

The Public Utility Regulatory Policies Act — the 1978 law that mandates that public utilities buy power produced by homeowners with windmills *(see box, p. 106)* — also applies to power produced by small-scale hydroelectric producers. According to Daniel Deudney of Worldwatch Institute, the law is having a "revolutionary impact on the economics of small-scale renewable power technologies. For the first time utility executives are being forced to compare the costs of such systems with those of large, centralized generating facilities."[26]

Other federal government incentives to get investors interested in rebuilding water power dams include low interest loans and special tax depreciation benefits. The transition to renewable energy sources has been a slow one. But the pace is likely to pick up when the price of imported oil begins to rise again.

[25] Quoted by Harris, *op. cit.*, p. 28.
[26] Deudney, *op. cit.*, p. 35.

119

Selected Bibliography

Books

Brown, Lester R., *Building a Sustainable Society,* Norton, 1981.
Eldridge, Frank R., *Wind Machines,* 2nd ed., Van Nostrand Reinhold, 1980.
Hayes, Denis, *Rays of Hope: The Transition to a Post-Petroleum World,* Norton, 1977.
McGuigan, Dermot, *Harnessing the Wind for Home Energy,* Garden Way, 1978.
Reynolds, John, *Windmills and Watermills,* Praeger, 1970.
Torrey, Volta, *Wind-Catchers: American Windmills of Yesterday and Tomorrow,* Stephen Greene Press, 1976.

Articles

Bourgeois, Jean-Louis, "Welcoming the Wind," *Natural History,* November 1980.
Harris, Michael, "Reinventing the Waterwheel," *Environmental Action,* June 1979.
Land, Thomas, "Wind Power," *Europe,* November-December 1980.
Minchinton, Walter, "Wind Power," *History Today,* March 1980.
O'Lone, Richard G., "Utilities Turn to Advanced Windmills," *Aviation Week & Space Technology,* March 23, 1981.
Smith, Norman, "Water Power," *History Today,* March 1980.
Smith, R. Jeffrey, "Wind Power Excites Utility Interest," *Science,* Feb. 15, 1980.

Reports and Studies

Clews, Henry, "Electric Power From the Wind," Enertech Corp., 1973.
Deudney, Daniel, "Rivers of Energy: The Hydropower Potential," Worldwatch Institute, June 1981.
Editorial Research Reports: "New Energy Sources," 1973 Vol. I, p. 185.
Flavin, Christopher, "Wind Power: A Turning Point," Worldwatch Institute, July 1981.
U.S. Department of Energy, "Monthly Energy Review," September 1981.
U.S. General Accounting Office, "Hydropower — An Energy Source Whose Time Has Come Again," Jan. 11, 1980.
Wentworth, Mary C. "Electric Utility Solar Energy Activities, 1980 Survey," Electric Power Research Institute, 1980.
"Wind Energy: An Introduction," American Wind Energy Association, 1980.

Cover by Staff Artist Robert Redding;
photograph p. 119 by Sally May Harris;
photograph p. 110 courtesy of the Department of Energy

NUCLEAR FUSION DEVELOPMENT

by

William Sweet

Sept. 12
1 9 8 0

Editor's Note: On Sept. 24, 1980, Congress cleared legislation to accelerate the development of fusion energy. The bill set a goal of having a commercial demonstration plant utilizing nuclear fusion operating by the year 2000, speeding up the Department of Energy's fusion research plans by 15 to 20 years. The bill did not authorize specific funding, but it required the secretary of energy to maintain a broadly based fusion research program; establish a national magnetic fusion center; initiate design of a fusion engineering test device, to be operational no later than 1990; and initiate as soon as practicable all activities required to meet the national goal of operating a nuclear fusion demonstration plant at the beginning of the 21st century. Congressional sponsors of the legislation say the Energy Department has done little to implement the provisions.

President Reagan's fiscal 1983 budget requests $444 million for magnetic fusion research (see p. 126), $10 million less than the fiscal 1982 amount.

NUCLEAR FUSION DEVELOPMENT

NUCLEAR fusion, the energy source of the sun and other stars, could provide the Earth with a virtually unlimited energy supply — if it can be harnessed effectively and economically. Since the beginning of fusion research three decades ago, it often has seemed that the harder scientists worked, the further away a viable fusion technology appeared. Pleading for time, fusion researchers would tell impatient critics that their work was the most challenging scientific program ever undertaken. But such claims only strengthened the impression that this was one confrontation between nature and man that nature was going to win.

In recent years, however, fusion researchers have shown a marked optimism. Experts in the field still are careful to point out that fusion research has so far failed to produce a single controlled fusion reaction. But now, in sharp contrast to earlier years, they go on to say that a successful demonstration of the technology is within sight. "We can't prove it yet, but we believe it," is how a Department of Energy planning official assessed the outlook in a recent interview.[1] Dr. Edward Kintner, who directs one of the department's two fusion energy programs, said that "the question of when we'll get workable fusion now is more a question of man than nature."[2]

Reflecting the new optimism about fusion's viability, government support for fusion research has grown significantly in recent years. Throughout the 1950s and 1960s, government funding for fusion research hovered at $20 million-$30 million. Not until the oil crisis of 1973-74 did appropriations begin to increase sharply. In fiscal year 1980, the authorization for the fusion program considered most promising — magnetic confinement fusion (see p. 126) — came to $350 million.

The House of Representatives, on Aug. 25, passed by a margin of 365-7 a bill authorizing $434.5 million in fiscal 1981 for research on magnetic confinement fusion. The bill, sponsored by Rep. Mike McCormack, D-Wash., chairman of the Subcommittee on Energy Research and Production and a longstanding advocate of fusion, stated that $20 billion should

---

[1] Dr. Michael Roberts, Director, Division of Planning and Projects, Office of Fusion Energy, interview with Editorial Research Reports, Aug. 21, 1980.
[2] Interview on Aug. 21, 1980.

be spent on fusion research over the next 20 years. "We are presently a society in transition from old energy sources to new ones," McCormack said, "and when we step across the line into the world of controlled magnetic fusion, as we can do before the year 2000, we will be taking the most important step in the history of mankind, as far as energy is concerned. It will be comparable to the first use of fire."[3] The Senate counterpart to the McCormack bill is sponsored by Sen. Paul E. Tsongas, D-Mass., a member of the Subcommittee on Energy Research and Development and vice chairman of the Subcommittee of Energy Conservation and Supply. The primary difference between the two bills is that McCormack's legislation calls for a fusion engineering test reactor to be built by 1987, while Tsongas' proposal sets a deadline of 1990.

The idea of building a test reactor came from the recommendations of a special Fusion Review Panel, which was established early this year by the Energy Research Advisory Board (ERAB) of the Department of Energy. In a draft version of its final report, issued on June 23, the panel concluded that U.S. taxpayers "are receiving their money's worth" from the government's fusion program.[4] The panel said that because of recent progress in the field, the United States is ready to embark on "exploration of the engineering feasibility of fusion." The engineering program envisaged by the panel would require "a doubling in the size of the present program ... in five to seven years" — roughly what the McCormack bill recommends.

Most members of the Fusion Review Panel work in high-technology fields[5] and might be expected to take a favorable view of fusion. Perhaps more surprising was the enthusiasm shown by a prominent scientist known for his skepticism about some advanced energy technologies — Dr. Thomas Cochrane of the Natural Resources Defense Council,[6] who attended the fusion panel meetings as a member of the Energy Research Advisory Board. Cochrane told the panel members that he "likes to bet on a winner, and [fusion] looks like a winner." In a recent interview, however, Cochrane raised questions about whether fusion research was taking an excessive amount of money away from other energy technologies — mainly solar

---

[3] *Congressional Record*, Aug. 25, 1980, H. 7661.

[4] "[Draft] Report of the Fusion Review Panel of the Energy Research Advisory Board," Department of Energy, June 23, 1980. The panel's final report was approved by the Energy Research Advisory Board in August.

[5] For example, the panel's chairman was S. J. Buchsbaum, vice president of Bell Laboratories.

[6] Cochrane has been a leading critic of the fast breeder reactor, an advanced fission reactor that produces plutonium, the key ingredient needed to build atomic bombs. The Natural Resources Defense Council, which has offices in New York, Palo Alto and Washington, D.C., is one of the nation's leading environmental organizations.

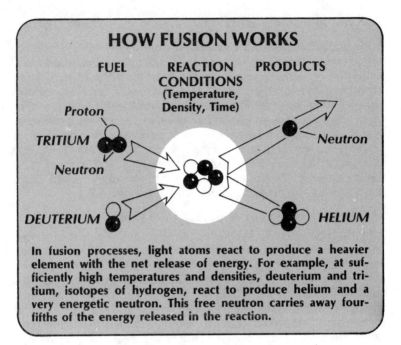

HOW FUSION WORKS

FUEL   **REACTION CONDITIONS** (Temperature, Density, Time)   **PRODUCTS**

Proton

TRITIUM

Neutron

Neutron

DEUTERIUM

HELIUM

In fusion processes, light atoms react to produce a heavier element with the net release of energy. For example, at sufficiently high temperatures and densities, deuterium and tritium, isotopes of hydrogen, react to produce helium and a very energetic neutron. This free neutron carries away four-fifths of the energy released in the reaction.

— which could do much more in the decades immediately ahead to help reduce U.S. reliance on foreign oil.

Fusion's Viability as an Energy Source

If fusion is to become a viable energy source, a way must be found to force atoms of hydrogen together in a controlled reaction. Intense heat and density are required to overcome the natural tendency of the nuclei to repel each other. When this is done successfully, an enormous amount of energy is released.

In the interior of the sun, powerful gravitational forces confine the nuclei at high densities, allowing fusion reactions to take place at temperatures of 15-20 million degrees centigrade. Comparable pressures cannot be duplicated on Earth, however. Artificial creation of sustained fusion reactions therefore requires much higher temperatures — over 100 million degrees centigrade. Confining and controlling matter at such high temperatures involves numerous scientific and engineering problems which are a long way from being solved.

The easiest kind of fusion process to start and sustain is the deuterium-tritium reaction, which results in creation of helium and highly energetic neutrons *(see box, above)*. It takes roughly 10,000 electron volts (10 KeV) to bring about a single deuterium-tritium reaction; the energy released in the reaction comes to 17.6 million electron volts (17.6 MeV). Deuterium and tritium are both heavy isotopes of hydrogen, one of the two constituent elements of water.

Since deuterium can be extracted from water relatively easily with current technology, its supply is considered virtually unlimited.[7] Tritium, on the other hand, must be "bred" from lithium, a less abundant substance. Moreover, tritium is radioactive and in gaseous form poses containment problems.[8] For these reasons, fusion would be more attractive as an energy source if it could be based on processes that avoided use of tritium, for example pure deuterium-deuterium reactions. But these reactions are even harder to produce and sustain, and their commercial application must be considered even more remote than the deuterium-tritium fusion process.

Magnetic Confinement Approach to Fusion

In the approach to fusion technology that is most advanced and most widely practiced, deuterium and tritium are confined in a doughnut- or torus-shaped tube that is surrounded by magnets and other equipment. At temperatures over 10,000 degrees centigrade, the light atoms dissociate into their electrically charged constituents — electrons and ions — forming a "plasma" *(see box, p. 131)*. Since charged particles tend to gyrate along magnetic field lines, the very hot plasma can be contained in the "magnetic bottle."[9]

Part of the heat required to bring the plasma up to fusion temperatures is provided by electric current, which is induced in the plasma tube, so that it flows the long way around the torus as horses race around a track. This "ohmic" or ordinary resistance heating can get the plasma temperature up to 10-20 million degrees centigrade. Supplementary means of heating have to be used to raise the plasma temperature up to the level needed for fusion reactions (over 100 million degrees centigrade). Generally this is accomplished by shooting deuterium atoms into the torus from accelerators located outside the magnets. Electrically neutral, the atoms pass freely through the magnetic fields and are stripped of their electrons only as they transfer their energy to the plasma upon collision with plasma particles.

During the first decade and a half of fusion research, scientists experimented with different shapes of magnetic tubing to find a design which would keep the slippery plasma in a stable formation. This proved to be extremely difficult, as many sources of instability were discovered. For example, under in-

[7] Deuterium-oxide, popularly known as "heavy water," is used as the moderator and coolant in the fission reactors that Canada produces and sells for generation of electricity. All other nuclear reactors currently built for commercial energy production use ordinary water as the coolant.

[8] Tritium is manufactured for use in hydrogen bombs in reactors at Savannah River, S.C. See Howard Morland, "Tritium: The New Genie," *The Progressive*, February 1979, pp. 20-24.

[9] Harold P. Furth, "Progress Toward a Tokamak Fusion Reactor," *Scientific American*, August 1979, pp. 50-61.

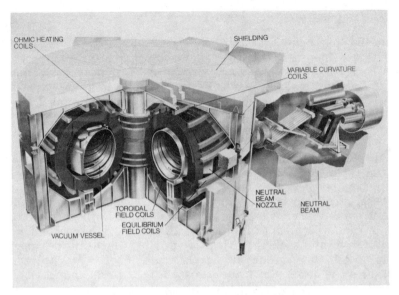

Model of the Princeton Large Torus Reactor

tense magnetic pressure the plasma would tend to bend, somewhat as a structure buckles under stress. In some cases impurities entering the plasma from the tube walls would lead to losses of heat. The reactor which proved most effective in dealing with these instabilities was invented in Russia in 1968 and is known as a "Tokamak." The name is a transliteration of three syllables: "to," for toroidal; "ka," for chamber; and "mak," for magnetic.

Most fusion experiments in the United States have been conducted at Princeton University's Plasma Physics Laboratory, where scientists began to work with table top-sized experimental reactors in the early 1950s. After the Russian breakthrough in 1968, the Princeton team concentrated on building tokamak reactors. A decade later, on July 4, 1978, the Princeton Large Torus Reactor attained for the first time a temperature of 60 million degrees and held it for 15 thousandths of a second. On May 1, 1980, the same machine achieved a temperature of 82 million degrees and held it for 25 thousands of a second.

The July 1978 experiment received considerable attention in the press, but its actual significance was somewhat ambiguous. Scientists at the Massachusetts Institute of Technology already had attained a higher combined measure of heat and confinement time and had come closer to "breakeven" — the point where fusion reactions generate as much energy as they are consuming.[10] But the Princeton scientists put special emphasis on the fact that plasma instabilities were much less

---

[10] See William D. Metz, "Report of Fusion Breakthrough Proves to Be a Media Event," *Science,* Sept. 1, 1978, pp. 792-794.

severe than expected, and their success in this respect appears to have helped win greater public confidence in their work. The Tokamak Fusion Test Reactor currently under construction at Princeton and expected to come into operation in 1981 will be more than twice as big as the currently operating Large Torus.

Inertial Confinement Produced by Lasers

An entirely different approach to fusion energy, called "inertial confinement," can most easily be understood as an effort to tap the energy released by tiny nuclear explosions.[11] The basic idea is to bombard milligram-sized pellets of deuterium and tritium with energy from lasers or light-ion beams — the "driver," as it is called in the professional jargon. Experimenters currently are working with glass pellets, which are designed to absorb energy from the beams and explode, which in turn causes the deuterium-tritium mixture to implode and fuse.[12]

The problems that must be solved in order to make inertial confinement work are no less formidable than the problems raised by tokamak experiments. Numerous lasers have to be able to focus energy on precisely the same point at precisely the same time, in pulses lasting a billionth of a second or less. The chamber in which the reactions take place has to withstand explosions equivalent to about 20 kilograms of high explosive, and as many as a hundred of these explosions would be taking place per second.

According to Dr. Robert L. Schriever, Deputy Director of the Energy Department's Office of Inertial Fusion, an energy breakeven experiment is hoped for in this field in about 1986; the new tokamak under construction at Princeton is supposed to attain breakeven in 1982.[13] Reflecting its more immediate promise, research on magnetic confinement fusion receives roughly twice as much funding as inertial confinement research — about $400 million versus $200 million.

The magnetic confinement program is managed by a branch of the Energy Department's research office, while the inertial confinement program is part of the department's defense division.[14] And while most research on magnetic confinement is done at universities such as Princeton and the Massachusetts Institute of Technology (MIT), the important experiments on inertial confinement are being conducted at the nation's weapons laboratories — Lawrence Livermore Laboratory in Liver-

---

[11] Like a hydrogen bomb, inertial confinement techniques rely on the inertia of light atoms rather than magnetic fields to confine the deuterium-tritium mixture during fusion reactions.

[12] See John L. Emmett, et al., "Fusion Power by Laser Implosion," *Scientific American,* June 1974, pp. 24-37.

[13] Interview with Schriever on Aug. 21, 1980.

[14] As the successor to the Atomic Energy Commission, the Department of Energy has responsibility for design and production of nuclear weapons.

more, Calif., Los Alamos Scientific Laboratory in Los Alamos, N.M., and Sandia Laboratories in Albuquerque, N.M.

In contrast to the magnetic confinement program, which has exclusively peaceful purposes, the laser-beam program was started primarily for military reasons and it continues to have important military applications. Schriever stressed that the program cannot result in production of a weapon; rather, its military objective is to test and evaluate the kinds of reactions which interest designers of advanced nuclear weapons.[15] Testifying before the Senate Subcommittee on Arms Control on April 28, 1980, Schriever said that the large facilities now being built to test inertial confinement techniques would be important in maintaining the viability of U.S. nuclear weapons laboratories in the event this country agrees to a comprehensive test ban treaty. Even though U.S. weapons designers no longer could test weapons because of treaty obligations, Schriever said, they could continue to simulate weapons effects in laser experiments and use the results to improve weapon design.

Early Experiments and Research

FUSION REACTIONS, like the fission reactions which power the nuclear reactors currently in operation, were first harnessed for human ends in connection with the production of atomic weapons. Fission chain reactions, in which heavy atoms like uranium or plutonium split, provide the explosive power of atomic bombs such as those that were dropped on Hiroshima and Nagasaki. In a hydrogen bomb, a fission explosive triggers fusion reactions.[16]

The initial impetus for development of fusion power came from Los Alamos, where a study program on thermonuclear reactions had been started soon after World War II, and from Lyman Spitzer Jr., an astronomer and student of interstellar gases at Princeton University. Spitzer sent some ideas for a fusion reactor to the Atomic Energy Commission, which led to a meeting of interested parties at AEC headquarters on May 11, 1951. The meeting resulted in the establishment at Princeton of a fusion research program called "Project Matterhorn." According to an informal history of the project, "[t]he

---

[15] Schriever said on Aug. 21 that pellet explosions might yield very intense neutron bursts, which would be especially interesting to people assessing neutron effects. For background on the development of neutron bombs or "enhanced radiation weapons," see "The Neutron Bomb and European Defense," *E.R.R.*, 1980 Vol. II, p. 581.

[16] See "Atomic Secrecy," *E.R.R.*, 1979 Vol. II, pp. 641-660.

work at hand seemed difficult, like the ascent of a mountain — and besides, Spitzer had pleasant memories of Switzerland."[17]

Physicist J. A. Van Allen, discoverer of the Van Allen radiation belts,[18] took a leave of absence from the University of Iowa to join the project in 1953, when the team was embarking on construction of its first experimental device. Edward A. Frieman, current director of research at the Department of Energy, also joined the project in 1953; Melvin B. Gottlieb, current director of the Princeton Plasma Physics Laboratory, joined the project in 1954.

During the 1950s and much of the 1960s, the Matterhorn team built and tested a series of reactors which for the most part were figure-eight shaped devices called "stellerators," modeled after the idea Spitzer described to the AEC in 1951. On the basis of these experiments, scientists acquired a much deeper understanding of plasma behavior, and engineers invented equipment which continues to be indispensable to fusion research. But the problems associated with taming the power of the sun proved to be even more formidable than expected, and most of the early machines failed to perform as well as had been hoped.

"In retrospect," one observer wrote, "it seems clear that although the key theoretical aspects of the problem were fairly well understood at the outset, the experimental difficulties were greatly underestimated."[19] According to Gottlieb, Spitzer's early theoretical ideas were "important and basically correct, but incomplete." By the early 1960s, Gottlieb recalled, a mood of pessimism had come over the program, as it was beginning to look as though a fusion reactor "just wasn't going to work."[20]

Russian Advances; Invention of Tokamak

Edward Teller, who generally is thought of as the "father" of the U.S. hydrogen bomb, played an important part in stimulating research into the peaceful applications of fusion energy. An even greater role was played by Russian physicist Andrei D. Sakharov, the "father" of the Soviet H-bomb. In an important paper written in 1948, Sakharov formulated the principles for the magnetic thermal isolation of high-temperature plasma.

---

[17] Earl C. Tanner, *Project Matterhorn: An Informal History,* Princeton University Plasma Physics Laboratory (1979 revision), p. 2.
[18] The Van Allen radiation belts are doughnut-shaped zones of charged particles, mainly protons and electrons, which are trapped in the earth's magnetic field. They surround the earth's equator starting several hundred miles out from the earth's surface. Their existence was first postulated by Van Allen in May 1958, on the basis of results from a Geiger counter that he had installed in the first U.S. satellite, Explorer I.
[19] Furth, *op. cit.,* p. 51.
[20] Interview on Aug. 25, 1980.

Plasma Physics

The term "plasma" probably is most widely understood as the word for the fluid part of blood, but in physics it refers to a state of matter in which the electrical bonds holding the electrons to atomic nuclei have been broken. This plasma or ionized gas consists of free electrons and free positrons in equal numbers and sometimes is called the "fourth state of matter," the other three being solids, liquids and gases.

The existence of a fourth state of matter first was postulated by a British scientist, Sir William Crookes, in 1879. Fifty years later, Fritz Georg Houtermanns and Robert d'Escourt Atkinson theorized that the source of energy in the stars results from the fusion of light nuclei in the stellar interiors. This discovery first was put to practical use after World War II with the invention of hydrogen bombs, which rely for their power on fusion or thermonuclear reactors.

Current evidence indicates that the largest portion of matter in the universe consists of inter-stellar and intergallactic plasma. Stars such as our sun consist of almost 100 percent ionized plasma.

His ideas significantly affected the course of Soviet fusion research. In 1950, together with Igor Tamm, another prominent Russian physicist, Sakharov outlined the conditions needed for controlled fusion reactions. From about 1948 to 1956, Sakharov was involved almost exclusively in weapons work, and his research was kept a closely guarded secret. But much of this work contributed to Russian advances in the peaceful applications of fusion.

An important milestone in the development of fusion technology was reached in 1958, when all the major nations participating in controlled thermonuclear research agreed to declassify their work. This was done in preparation for the Second International Atoms for Peace Conference, which took place in Geneva that year, pursuant to the Atoms for Peace Program which President Eisenhower launched in 1953. U.S. scientists naturally were eager to learn what the Russians had accomplished. As it turned out, that interest proved well justified, for it was in the Soviet Union that the most successful fusion reactor — the tokamak — was invented.

The invention of the tokamak in 1968 at the I. V. Kurchatov Institute of Atomic Energy in Moscow attracted wide attention and provided a new stimulus for fusion research in many parts of the world. Added impetus came when the Organization of Petroleum Exporting Countries (OPEC) began to drive up oil prices beginning in 1973. While nobody saw fusion as a technology that would help reduce dependence on oil any time soon, the oil crisis highlighted the dangers of relying too much on non-renewable fuels and generated strong sentiment — especially in the advanced industrial countries — for development of new energy sources.

There are now four major tokamak construction programs under way in the advanced industrial countries: Princeton University's Tokamak Fusion Test Reactor *(see p. 128)*; the "T-15" reactor in Russia; a "JT-60" device in Japan; and the Joint European Torus, in Culham, England, a project which is being funded by member countries of the European Economic Community.

New Restrictions on Information Exchanges

Scientific exchanges between the United States, Europe and Japan are growing rapidly, and researchers expect these exchanges to yield great benefits. But ever since the Carter administration imposed restrictions on scientific exchanges with the Soviet Union, in response to Russia's invasion of Afghanistan last December, contacts between U.S. and Russian fusion researchers have been sharply curtailed. Melvin B. Gottlieb, who soon will retire as head of the Princeton laboratories, said in an interview on Aug. 25 that 1980 will be the first year since 1958 that he has not travelled to the Soviet Union to meet with Russian researchers. Gottlieb and other people connected with the U.S. fusion effort do not quarrel with the reasons for restricting scientific exchanges with Russia, but they leave little doubt that their work would be easier if the free interchange of ideas and information could resume once political issues are resolved.

Regardless of how U.S.-Soviet relations evolve, members of the U.S. program believe that they have taken the lead in world fusion research, and they would like to keep it. The United States now accounts for roughly one-third of total world expenditures on fusion, and people like Gottlieb strongly believe that the "United States doesn't want to lose the lead in fusion, as it has in fission."[21] The luxurious office building into which Gottlieb's staff recently moved, and the impressive building being built next door to house the tokamak test reactor, leave little doubt that the Princeton program is meant to be one of America's most highly prestigious efforts in the realm of advanced technological research.

---

[21] Gottlieb was referring to the Carter administration's effort to curtail the U.S. breeder reactor program in the interest of curbing nuclear proliferation. Like many other scientists and nuclear industry executives, Gottlieb believes that restrictions on breeder reactors have only enabled France and the Soviet Union to take the lead in this field.

Outlook for Commercialization

T HE immediate objectives of fusion programs are to achieve "ignition" of a deuterium-tritium mixture and to attain energy "breakeven." If these goals are reached in the years immediately ahead, as researchers hope, the scientific feasibility of fusion for energy production will be demonstrated. Very significant problems still will have to be solved, however, before fusion will be demonstrated as an economically viable method of generating electricity.

Gottlieb described several areas in which engineering efforts will have to be concentrated. Heat losses from the plasma to the reactor walls must be reduced, and care must be taken to prevent "hot spots" from developing at points along the walls. Large super-conducting magnetic coils — electric magnets that are kept at very low temperatures — eventually will have to be built to attain the confinement times and heat required. A "blanket" must be developed for the reactor which can both transfer heat from the vessel to power generators and breed tritium from lithium. In designing the blanket and containment structure, close attention will have to be paid to protection of employees and the public from releases of radioactive tritium. Finally, remote handling equipment must be designed so that radioactive parts can be maintained and replaced without hazard to human health.

Fusion researchers are cautious about predicting how fast they will make progress in solving such problems. Throughout the scientific and business communities, there is a sharp awareness of the unforeseen difficulties which engineers have had in making fission reactors economic, reliable and publicly accepted means of energy production. Considering that the problems connected with fusion are much more challenging, and in view of the way that fission reactors were over-sold as a source of electricity "too cheap to meter," no one wants to raise excessively high expectations about fusion.

Fusion's Advantages Vis-à-Vis Fission

It was relatively easy to adapt fission energy to peaceful purposes, basically because nuclear power plants of the kind currently operating are simply more elaborate versions of the machines that were built to provide plutonium for the first atomic bombs. In its essential features, a commercial nuclear power plant is the same device which Enrico Fermi's team built at the University of Chicago in 1942 to create the world's first self-sustaining nuclear chain reaction. The adaption of fusion energy to peaceful purposes has been immensely more difficult, primarily because the method of creating a fusion reaction in the H-bomb, in which a fission explosive is used to provide the requisite heat and pressure, cannot be directly applied to power reactors. Harnessing fusion energy for power production has required scientists and engineers to develop entirely new equipment. If that equipment eventually is made to work, fusion could have some important advantages over fission.

"The United States doesn't want to lose the lead in fusion, as it has in fission."

—Melvin B. Gottlieb,
Director of the Princeton
Plasma Physics Laboratory

One of fusion's advantages is its immunity to the kind of runaway reaction which can lead to the melt-down of a fission reactor and the release of noxious radioactive substances. The total amount of plasma contained even in a very powerful tokamak reactor would be very small, on the order of a few grams. Therefore, even if the fusion reaction got out of control, the heat contained in the plasma would be rapidly absorbed by the much larger containment vessel. Stopping a fusion reaction, according to one expert, would be "like throwing a bucket of water on a match."[22]

In the event a fusion containment vessel ruptured, the gases released would be much less noxious than those which would escape following the melt-down of a fission reactor. Fusion does produce some radioactivity, however. Containment of tritium in normal and abnormal fusion reactor operations poses serious engineering problems, even though tritium's radioactivity is short-lived and is rapidly diluted as the gas disperses in the air. In addition, neutron bombardment of the reactor vessel during fusion reactions makes the walls of the vessel so highly radioactive that a way must be found of repairing and replacing them by remote control.

Another advantage of fusion over fission is its less intimate link with production of atomic weapons. The commercial nuclear power plants currently operating produce plutonium, the essential substance required to manufacture atomic bombs, and

---

[22] Interview with Dr. Edward Kintner, Aug. 21, 1980.

there has been mounting concern about whether the dissemination of these plants will enable many non-nuclear countries to build nuclear weapons.[23] Tokamak reactors, on the other hand, would not be an important source of materials needed for construction of hydrogen bombs.[24] Only the general scientific knowledge associated with tokamak design would be of use to H-bomb manufacturers.

Military Applications of the Technology

Some arms control experts have worried that the wide dissemination of knowledge about fusion processes eventually will make it easier for non-nuclear countries to design and build hydrogen bombs. People trained to work on experimental fusion reactors obviously would make good recruits for a weapons program, just as veterans of the U.S. and Russian hydrogen bomb programs made important contributions to peaceful fusion research. In addition, there is some concern that so-called "hybrid" tokamak reactors — a combination of fusion technology with the fission breeder concept — could be used to manufacture fissionable material for atomic bombs.[25]

With the invention and development of lasers in the early 1960s, an area of fusion research opened up that had much closer relevance to the design of hydrogen bombs *(see p. 128)*. Research on laser- and beam-driven fusion reactions only got going in earnest during the 1970s. But some arms control experts already have suggested that laser research should be restricted because of its applicability to weapon design. Writing in *The Bulletin of the Atomic Scientists*, John P. Holdren, a physicist at the University of California, pointed out that an "important barrier to the spread of fusion weapons has been the lack of access to certain technological insights, and the spread of inertial confinement approaches to fusion may spread that limiting ingredient."[26]

Many scientists instinctively resist the idea of trying to curtail the spread of information, and certainly most of the scientists associated with fusion research believe that the benefits of openness have far exceeded the dangers. As far as the risks are concerned, construction of a hydrogen bomb poses formidable engineering problems, even when there is available personnel well-trained in the basic science of fusion. Moreover, a hydrogen bomb can be built only after a country has

---

[23] See "Nuclear Proliferation," *E.R.R.*, 1978 Vol. I, pp. 201-220.
[24] Tokamaks probably would produce tritium, an important H-bomb ingredient, but there are other ways of breeding tritium which are easier and cheaper.
[25] In a hybrid reactor, neutrons from the fusion reactions would be used to breed plutonium from uranium in the reactor blanket.
[26] John P. Holdren, "Fusion Power and Nuclear Weapons: A Significant Link?" *The Bulletin of the Atomic Scientists*, March 1978, p. 5.

constructed an atomic bomb, and there is some question as to how much more danger results from addition of hydrogen bombs to arsenals that already include atomic bombs.

As for the benefits associated with openness, they appear to be very great indeed. Since fusion research is confined for all practical purposes to individual programs funded by the governments of the leading industrial countries, imposition of tight classification requirements on fusion research would sharply curtail the amount of information available to members of each program. Each program is a resource for every other program, and there is no hope of obtaining comparable resources from private universities or private industry in the foreseeable future.

Debate Over Potential Benefits of Fusion

Since fusion is not expected to reach commercial fruition for roughly 50 years, private businesses have not been interested in putting large sums of money into fusion research. One private company, General Atomics of La Jolla, California, has been involved in fusion research since the early 1950s, but since 1966 its fusion work has been funded almost exclusively by the Department of Energy and its predecessor organizations. Some power plant manufacturers and utilities follow advances in the field closely, and some businesses assign personnel to fusion research, but the prevailing attitude in the private sector is one of caution.

In an article on fusion which the Electric Power Research Institute published in 1977, it was pointed out that a 1,500 megawatt fusion reactor might require as much as 50,000 tons of type-316 stainless steel. "At today's prices," EPRI commented, "this amount would, by itself, exceed the total cost of a present-day fossil or fission power plant of equivalent output."[27] Considering that these steel walls might have to be replaced as often as every two years, the costs of a commercial tokamak-style reactor could turn out to be astronomically high.

Operating reactors at lower temperatures would help extend wall life, if this proves to be feasible, and it may be possible to develop materials for the walls that would be much more resistant to damage from neutron bombardment. Even so, it must be remembered that these walls are just one component of the reactor, and not the most expensive one at that. The single most expensive part of the reactors will be the superconducting magnets which still must be developed.

[27] "Capturing a Star: Controlled Fusion Power," *EPRI Journal,* December 1977, p. 13.

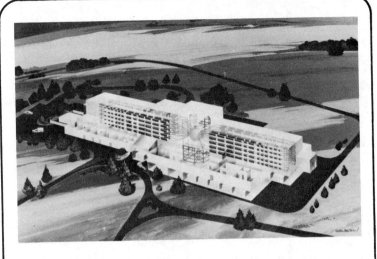

The world's largest laser fusion facility, called Shiva, is located at Lawrence Livermore Laboratory in California. Named after the multi-armed Indian god, the Shiva system consists of 20 laser tubes and a target chamber where the beams converge on pellets of fusion fuel. This system, as it currently operates, is located in the west wing of the building depicted in the drawing above.

A new wing, which will house the "Shiva Nova" system, currently is under construction. After the new wing is operating, the existing system will be shut down and rebuilt. Eventually the lasers in both wings will be operated as one system — bringing more than 200 trillion watts of energy to bear on the fusion pellets in the target chamber. Scientists hope Shiva Nova eventually will produce 20 times as much energy as that supplied by the laser beams to the pellets.

Proponents of fusion argue that the research has applications which go far beyond energy production. Even if the path to a commercially viable reactor proves to be more difficult than currently foreseen, it is implied, fusion research may pay off in other areas. Kintner, the manager of the Energy Department's magnetic confinement program, pointed out that many of the components designed for the tokamak reactors — vacuum tubes, super-magnetic conductors, etc. — will be of wide use in many parts of U.S. industry. Gottlieb emphasized that the Princeton program is "primarily a physics laboratory in which we investigate the interaction of plasmas and magnetic fields." The tokamak reactor is first and foremost an experimental device for use in basic research, he said, and only secondarily an energy producer. The engineers currently outnumber the scientists in the Princeton program, but Gottlieb said this is only because the experimental equipment is so complex.

Reflecting their new optimism, fusion researchers recently have taken measures to publicize their work. Earlier this year, the Plasma Fusion Center at the Massachusetts Institute of Technology started to publish the *Journal of Fusion Energy.* In August 1979, the Fusion Energy Foundation opened operations in Gaithersburg, Md., with the objective of promoting fusion research. According to its director, Dr. Steven Dean, a former assistant to Kintner at the Energy Department, the foundation has 17 corporate members and four non-voting affiliates.

"When we step across the line into the world of controlled magnetic fusion, as we can do before the year 2000, we will be taking the most important step in the history of mankind, as far as energy is concerned.

—Rep. Mike McCormack, D-Wash.

Some critics of fusion argue that the technology may already have been over-sold. Writing in *Science* magazine in 1976, journalist William D. Metz asserted that "a gap is developing between what the fusion program appears to promise and what it is most likely to deliver."

> Fusion power would not be infinitely abundant [Metz said] because the present reactors rely on a fuel cycle that must have lithium, and lithium is not particularly more abundant than the uranium-238 that would fuel the breeder reactor. Fusion power would not be free of radioactive gases because the lithium would be used to breed tritium — a gas more benign than fission products, but one that is devilishly hard to contain. The fusion reactor would not be free of waste disposal problems ... because ... a neutron flux would make all the structural materials of the reactor intensely radioactive — one estimate of the waste disposal problem is at least 250 tons of material every year from each reactor.[28]

Metz went on to suggest that alternative fuels deserved more emphasis and that the laser devices, which can be built in small modular units, may eventually prove more useful to utilities than the very large, enormously complex tokamak-type reactors.

---

[28] William D. Metz, "Fusion Research: What Is the Program Buying the Country?" *Science,* June 25, 1976, p. 1320.

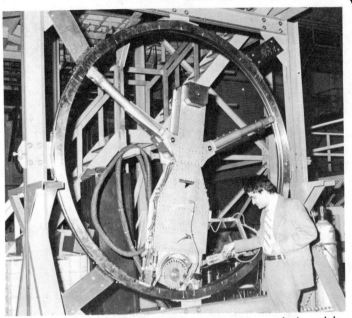

The remote cutter device pictured above was designed by
Grumman Aerospace Corporation to repair the Tokamak Fusion
Test Reactor currently under construction at Princeton Uni-
versify (see p. 128). The cutter can be folded up and inserted
through a porthole into the reactor torus by means of a remote
manipulator. Inside the reactor it is unfolded and used to cut
out cylindrical segments of the torus from the inside. Remotely
controlled cranes lift out the damaged segment, together with
the magnet segments which surround it. The remote welding
device that would be used to attach the new segment to the
torus is under development. The remote cutter, which weighs
less than 400 pounds, was first tested successfully on April
22, 1980.

Robert L. Schriever, the deputy director of the Energy
Department's laser fusion program, described fusion as a "great
adventure," in which "truly astounding and marvelous things
are being learned." Understandably, he is especially enthu-
siastic about his own branch of the program, which he described
as "young," "crisp," and "audacious." To hear him is to believe
him, and yet there are other people who make equally convinc-
ing and passionate claims for competing technologies. For the
untrained lay-person, it is never easy to evaluate such matters,
and in the coming years the choices that fusion involves are
likely to be extremely difficult.

Selected Bibliography

Books

Nuclear Energy Policy Study Group, *Nuclear Power Issues and Choices,* Ballinger, 1977.

Articles

"Capturing a Star: Controlled Fusion Power," *EPRI Journal* (published by the Electric Power Research Institute), December 1977.

Emmett, John L., et al., "Fusion Power by Laser Implosion," *Scientific American,* June 1974.

Furth, Harold P., "Progress Toward a Tokamak Fusion Reactor," *Scientific American,* August 1979.

Metz, William D., "Fusion Research," *Science,* June 25, July 2, and July 23, 1976.

Robinson, Arthur L., "Academics Victims in Fusion Politics Tangle," *Science,* Dec. 21, 1979.

Rose, David J., and Michael Feirtag, "The Prospects for Fusion," *Technology Review,* December 1976.

Reports and Studies

Editorial Research Reports: "Atomic Secrecy," 1979 Vol. II, p. 641; "New Energy Sources," 1973 Vol. I, p. 185; "The Neutron Bomb and European Defense," 1980 Vol. II, p. 581.

"Laser Fusion: A Collection of Articles from Energy and Technology Review," Lawrence Livermore Laboratory, September 1979.

"Laser Fusion Program at Los Alamos," Los Alamos Scientific Laboratory, 1979.

"Particle Beam Fusion," Sandia Laboratories, 1980.

Tanner, Earl C., "Project Matterhorn: An Informal History," Princeton University Plasma Physics Laboratory, September 1977 (revised 1979).

"The Princeton University Plasma Physics Laboratory: An Overview," Princeton University, October 1979.

"The Report of the Fusion Review Panel of the Energy Research Advisory Board (ERAB)," Department of Energy, June 23, 1980 (draft version).

Synthetic Fuels

by

Marc Leepson

Aug. 31
1 9 7 9

Editor's Note: On June 30, 1980, President Carter signed into law a multibillion-dollar aid package for development of synthetic fuels. The legislation authorized $20 billion to be allocated to private industry by a Synthetic Fuels Corporation. The corporation was directed to stimulate the production of 2 million barrels a day of synfuels by 1992 by investing federal funds through loans, loan guarantees and price guarantees.

Even before the corporation became operational in January 1982, the administration approved $3.5 billion in loans and loan guarantees for three synfuel projects; one to make natural gas from coal in North Dakota and two to make oil from shale rock in Colorado. Eleven additional projects are now being considered by the corporation — six in the South and five in the West. Two factors cloud the corporation's future: likely cuts in funding by the Reagan administration, and the recent decline in world oil prices, which has led some observers to question the need for a multibillion-dollar government effort to develop alternative fuels.

SYNTHETIC FUELS

SYNTHETIC FUELS have never been an important part of the U.S. energy picture. But what some are calling "synthetic fuels fever" broke out in Washington this summer. Spurred primarily by the long gas station lines that had appeared across the nation, the House of Representatives passed a bill in June to boost U.S. production of synthetic fuels — primarily oil derived from coal and shale rock. Then on July 15, President Carter called for an $88-billion, 10-year program to produce 2.5 million barrels of synthetic oil a day by 1990. If approved by Congress, Carter's program will put in motion one of the most extensive technological and financial ventures ever undertaken. His energy proposals are likely to occupy much of Congress' time for the rest of the year. Senate action on the House-passed bill is due to resume soon after the lawmakers return from summer recess on Sept. 5.

Building a synthetic fuels industry today has been compared to the two other huge American scientific undertakings of the century: development of the atomic bomb and putting astronauts on the moon. It faces some large hurdles; foremost is the matter of financing. Analysts say that each major synthetic fuel plant will cost about $1 billion to build. Few companies could undertake such a project without generous government financial support. And no matter how high the per barrel price of imported oil is at any time, the projected price of a barrel of synthetic oil is substantially higher.[1] There are also serious questions about the environmental effects of a large number of synthetic fuel plants. "These are not nice plants," Robert Hanfling of the U.S. Department of Energy said recently. "These are big, dirty plants. Everybody wants these plants — but wants them someplace else."[2]

Looming above the synthetic fuels question is the fact that the world is slowly but surely running out of oil. As the situation has been summed up by *Business Week,* the world is now entering "the end of the petroleum age." "For the first time since oil became a major source of energy," the magazine said in a recent special report, "the world's factories, fleets of automobiles, and other users are burning oil at a faster rate than it is being

[1] The average price of imported oil today is about $20 a barrel. U.S. Department of Energy researchers estimate that synthetic oil would cost about $25 to $40 a barrel.
[2] Quoted in *The Wall Street Journal,* July 12, 1979.

discovered."[3] According to the Department of Energy's 1978 annual report to Congress, the non-communist world consumed some 20.6 billion barrels of oil last year. But only about 14 billion barrels are being discovered annually. Although there are hundreds of billions of barrels in reserves,[4] the fact remains that most experts see the inevitable end in the decades ahead. It is agreed that the United States now needs a long-range energy plan based on sources other than oil.

Many hope that synthetic fuels will answer the nation's future energy needs. These fuels do not fit the strict definition of the word "synthetic." They are not laboratory-made artificial products. The fuels that are widely referred to as synthetics (or synfuels) are quite similar to crude oil and natural gas. Like crude oil, synthetic oils contain carbon and hydrogen and can be modified from their raw forms into more usable ones. Unlike crude oil, synthetics must undergo a process of synthesis to make them usable.

The synthetics that have received the most attention are oil and gas derived from coal, and oil extracted from shale rock. Government geologists estimate that 1.8 trillion barrels of oil are encased in shale rock beneath 11 million acres in Colorado, Utah and Wyoming *(see map, p. 147)* That amount is several times greater than all the proved reserves of crude petroleum on earth. About 80 billion barrels of that oil, scientists say, are readily recoverable. Shale oil has been recognized as a possible source of oil for over a century, but environmental problems and high costs have held back large-scale development.

Another potentially important synthetic fuel is called biomass — farm and forest products, residues and municipal solid wastes that are converted through fermentation into fuels.[5] The best known example is "gasohol," a mixture of gasoline and alcohol for use in automobiles. Another fuel mentioned as a "synthetic" is nuclear fusion, which as a fuel process currently exists only in theory but, if developed someday, would be virtually free of hazardous wastes and radiation that now plague the nuclear power industry *(see p. 155)*.[6] Still other fuels in the synthetic category are heavy crude, a type of oil that does not flow easily and is extremely difficult to extract, and oil encased in tar sands, most of which is in Canada.

Most of these synfuels have been in production in a small way at different times in the last 100 years. But the discovery in the

---

[3] "The Oil Crisis is Real This Time," *Business Week*, July 30, 1979, p. 45.
[4] The *Oil and Gas Journal* in its issue of Dec. 25, 1978, estimated that the non-communist world's total "proved" oil reserves were some 547.6 billion barrels. Proved reserves pertain to oil recoverable from known reservoirs under existing economic and operating conditions.
[5] See "Solid Waste Technology," *E.R.R.*, 1974 Vol. II, pp. 641-660.
[6] See "Nuclear Waste Disposal," *E.R.R.*, 1976 Vol. II, pp. 883-906, and "Determining Radiation Dangers," *E.R.R.*, 1979 Vol. II, pp. 561-580.

1930s of vast oil deposits, first in the United States and then in the Middle East, shoved the development of these fuels to the background. Today there is virtually no commercial synfuel production in this country.

Lawmakers' Past Action, Current Response

"To give us energy security, I am asking for the most massive peacetime commitment of funds and resources in our nation's history to develop America's own alternative sources of fuel," President Carter said in his televised address to the nation on July 15. With that announcement, he added to the synfuel fever in Washington. Three weeks earlier, the House overwhelmingly passed a multibillion-dollar synfuels bill. That vote was not the first on synfuels. As part of the nation's push to energy independence in 1975, President Ford's proposal to provide billions of dollars in government loans for the creation of a large synfuels industry was approved by the Senate. But the House did not act on the measure and it died.

The next year the House again rejected legislation authorizing federal loan guarantees and price supports for the development of synthetic fuels. A coalition of fiscal conservatives and environmentally sensitive liberals combined to defeat the measure, but by only one vote.[7] Soon after the House voted, the General Accounting Office (GAO) released a report[8] that posed serious questions about making a national commitment to synthetic fuel production. The report concluded that the most cost-effective way to produce energy would be to conserve it rather than to make a massive commitment to synfuels.

In 1977, the House and Senate authorized guarantees of up to $6 billion in loans to companies willing to undertake synfuels demonstration projects. But that authorization was part of an energy research bill that President Carter vetoed and Congress was unable to override. In 1978, the House passed a measure authorizing $75 million for construction of a plant to make liquid fuel from coal, but the Senate never took action on the bill.

On June 26 of this year, the House voted 368-25 to approve a multi-billion-dollar program to create a U.S. synfuels industry. The vote came at a time when oil shortages were causing long lines at gasoline stations in many parts of the country. The bill had bipartisan support and was approved quickly without recording, by name, how the representatives voted. "The public wants bold action. They do not want timid action," said Majority Leader Jim Wright, D-Texas. "The time has come for us to

---

[7] See 1976 *Congressional Quarterly Almanac*, p. 174.
[8] "An Evaluation of Proposed Federal Assistance to Financing Commercialization of Emerging Energy Technologies," Aug. 24, 1976.

do everything within our power to break the stranglehold upon this country that the foreign nations are asserting."

The House-passed bill, which Carter endorsed, calls on the government to encourage production of the equivalent of 500,000 barrels a day of synfuels by 1985, with the additional goal of 2 million barrels a day by 1990. The object is to ease U.S. dependence on foreign oil. The nation today imports about 7.7 million barrels of oil a day, about 43 percent of its consumption. The House bill would authorize $3 billion for price supports for synfuels.

Carter, announcing his synfuels program after the House passed its synfuels bill, requested the creation of an Energy Security Corporation, an independent, government-sponsored enterprise which would direct an $88 billion federal investment in the U.S. synfuels industry. The sole objective of the proposed corporation would be the development of domestic production of the several varieties of synfuels. Carter also asked Congress to establish an Energy Mobilization Board with authority to override federal, state or local environmental laws if they interfere with completion of high-priority synfuels projects.

The president said July 15, "We will protect the environment. But when this nation critically needs a refinery or pipeline, we will build it." Since then, congressional committees have split over how much authority the board should have to override environmental laws. The House Commerce Committee has voted to give the board the power that Carter requested, but the Senate Energy and House Interior committees have declined to do so.[9] Congressional observers currently are saying that a synfuels bill probably will emerge from Congress, but chances are it will be scaled down greatly from the mammoth program Carter envisions.

Commercial Activity in Colorado Oil Shale

While no large-scale commercial synfuels plants exist today in this country, a number of prototype plants are in operation or under construction. In January, a subsidiary of Occidental Petroleum Corp. began work on a $1 billion oil shale processing plant on a 5,120-acre tract in Colorado leased from the federal government. Occidental plans to begin commercial production of 50,000 barrels of shale oil a day in 1986. Occidental's chairman, Armand Hammer, is a long-time backer of oil shale production and is optimistic about the plant's future. "If the government would take the proper actions now," Hammer said recently, "we could get 2 million barrels a day by 1990."[10]

---

[9] See *Congressional Quarterly Weekly Report*, Aug. 4, 1979, p. 1581.
[10] Quoted in *The Wall Street Journal*, Aug 16, 1979.

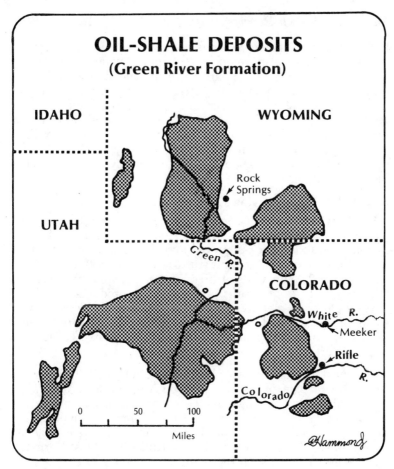

OIL-SHALE DEPOSITS
(Green River Formation)

IDAHO

WYOMING

UTAH

Rock
Springs

Green R.

COLORADO

White R.

Meeker

Rifle

R.

Colorado

0   50   100

Miles

Hammond

Occidental engineers predict that the site, between the towns of Meeker and Rifle, could eventually produce as much as 1.5 billion barrels of oil.

Since 1972 Occidental has been developing an *in situ* — in place — oil-shale recovery process that differs from conventional recovery methods. Instead of digging out the shale-bearing rock and crushing it, an explosive is detonated within the underground shale deposit, setting off a slow-burning fire that releases the oil. The liquid then drips into a cavity below the heated deposit and is pumped out. A 100-mile long pipeline will be built to move the oil into the distribution system for Midwestern refineries. At least three other shale-oil synfuel plants in Colorado are in planning stages.[11]

The government and private industry have been working to improve the methods for converting coal to liquids for many years. Despite high costs that have prevented commercial mar-

[11] Gulf/Standard (Ind.) has a pilot plant under construction and another planned. Union Oil and Arco/Tosco also have plans for oil shale development.

keting, the U.S. Energy Department philosophy has been to support commercial development of coal liquefaction as insurance against future oil embargoes. In partnership with private industry, the agency has supported construction of pilot and demonstration plants. The government built a pilot plant in 1974 at Ft. Lewis, Wash., to test a type of coal liquefaction process called Solvent Refined Coal (SRC-II). The $50 million plant is operated by Pittsburg and Midway, a Gulf Oil Co. subsidiary, and processes about 30 tons a day of coal into 100 barrels of oil.

New Plants for Removing Oil From Coal

The SRC-II process not only converts coal into liquid fuel, but also produces gases such as methane, naphtha, propane, butane and ethane. In the SRC-II process, crushed coal is mixed with a solvent and hydrogen at temperatures of up to 850 degrees and under some 2,000 pounds of pressure per square inch. Much of the energy in the process is provided by byproducts, including natural gas from methane, undissolved coal and the ash. Some 4,500 barrels of fuel from the Ft. Lewis plant were burned successfully in a test conducted by the Consolidated Edison Co. of New York.

The next demonstration plant to test the SRC-II process is scheduled for construction in 1981 in Morgantown, W.Va., by Gulf's subsidiary and the federal government. Upon completion in 1985 at an estimated cost of $700 million, the plant is expected to use about 6,000 tons of coal and produce some 20,000 barrels of coal liquids a day. The governments of West Germany and Japan will help with the plant's financing and share in the technical knowledge. "After two years of operation, we know the technology," said William C. King, a Gulf official. "It is feasible to convert coal to a range of liquid and gaseous fuels."[12]

Another $700 million coal liquefaction plant is due to be built — this one near coal fields at Owensboro, Ky. It will produce a clean-burning solid (rather than a liquid) substitute fuel using the SRC-II process. The plant will be similar in size and capacity to the Morgantown plant, and will be sponsored by the U.S. government and the Southern Co., a group of utilities in the South.

A liquefaction process called Exxon Donor Solvent (EDS), developed by the Exxon Corp. in the late 1960s, will be tested further under a $12 million contract awarded by the Department of Energy in 1976. The department and a consortium headed by Exxon have agreed to build a 250-ton-per-day pilot plant at Baytown, Texas. Construction began in May 1978 and the plant

---

[12] Quoted in *The New York Times*, Aug. 5, 1979.

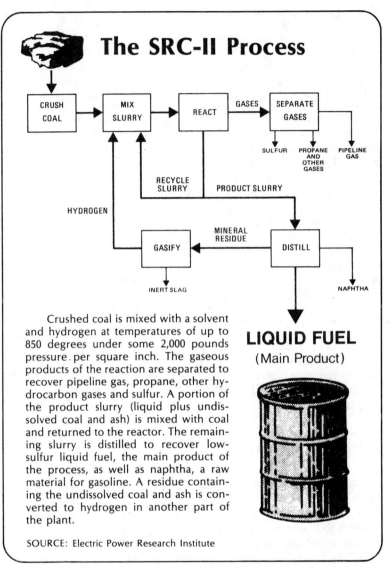

The SRC-II Process

CRUSH COAL → MIX SLURRY → REACT → GASES → SEPARATE GASES

SEPARATE GASES → SULFUR / PROPANE AND OTHER GASES / PIPELINE GAS

RECYCLE SLURRY

PRODUCT SLURRY

HYDROGEN

GASIFY ← MINERAL RESIDUE ← DISTILL

INERT SLAG

NAPHTHA

LIQUID FUEL
(Main Product)

Crushed coal is mixed with a solvent and hydrogen at temperatures of up to 850 degrees under some 2,000 pounds pressure per square inch. The gaseous products of the reaction are separated to recover pipeline gas, propane, other hydrocarbon gases and sulfur. A portion of the product slurry (liquid plus undissolved coal and ash) is mixed with coal and returned to the reactor. The remaining slurry is distilled to recover low-sulfur liquid fuel, the main product of the process, as well as naphtha, a raw material for gasoline. A residue containing the undissolved coal and ash is converted to hydrogen in another part of the plant.

SOURCE: Electric Power Research Institute

is due to be completed by mid-1982. The Exxon process produces low-sulfur oil and liquids that can be used for blending with gasoline. *The Energy Daily* newsletter in Washington, D.C., characterized Exxon's coal liquefaction as one of the most "promising new processes in the whole field of synthetic fuel development — perhaps the most elegant, conservatively designed and hydrogen-efficient of all the direct liquefaction processes."[13]

Starting up earlier than the others is a coal liquefaction plant at Catlettsburg, Ky., due to begin operations Jan. 1, 1980. The

[13] Issue of Aug. 8, 1979, p. 5.

pilot 600-ton-per-day plant, operated by a consortium led by Ashland Oil Co., will cost about $296 million and is expected to produce up to 2,000 barrels of an industrial fuel oil a day. It will use the H-Coal process, which was developed by Hydrocarbon Research Co. 10 years ago at small pilot plants at Trenton, N.J.

Still another method of extracting oil from coal involves breaking down the coal into carbon monoxide and hydrogen gases and recombining them in liquid form. Called the Sasol Synthol process, it is based on a German (Fischer-Tropsh) method used during World War II *(see p. 152)*. The South Africa Coal, Oil and Gas Corp. has been producing oil from coal with the Sasol method since 1955 at a plant 60 miles north of Johannesburg. South Africa, like the United States, is rich in coal. Unlike this country, South Africa has no oil of its own. Sasol currently produces about 20,000 barrels of synthetic oil a day — less than 10 percent of the nation's needs. Sasol, backed by the South African government, is planning to build two new plants. The expanded Sasol operation will produce some 100,000 barrels of hydrocarbon liquids a day by the early 1980s, filling about half of South Africa's growing needs.

The development of coal gasification techniques has lagged behind coal liquefaction and oil shale. To convert coal to synthetic gas, coal is fed with steam and oxygen into a high-temperature pressurized reactor. The result is a product called low-Btu gas, which has a lower heat value than natural gas but can be used as boiler fuel. A process called methanation can be used to convert low-Btu gas into a substance that has approximately the same heat content as natural gas.

A coalition of pipeline companies and a gas company called Great Plains Gasification Associates wants to market coal gas from a proposed $1.5 billion project in Mercer, N.D. As envisioned, it would produce 125 million cubic feet of gas a day — enough to heat some 365,000 homes a year. But the coalition has run into problems with the Federal Energy Regulatory Commission; the coalition wants customers to pay rates that cover the cost of the facility, but the commission has not agreed to permit this.

Development of Synthetic Fuels

NEARLY ALL the synthetic fuels under discussion today have been produced and used in the past. Oil shale, for example, has been a source of oil and combustible gas in fuel-short regions of the world since the 1840s. Deposits have been mined on a modest scale in Australia, Estonia, France, Man-

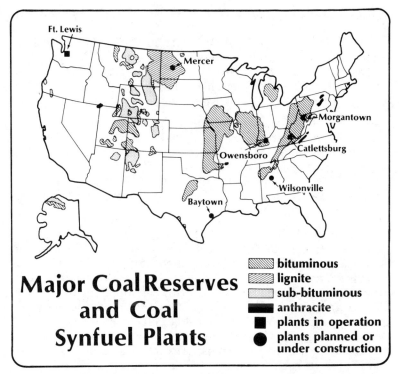

Major Coal Reserves and Coal Synfuel Plants

- Ft. Lewis
- Mercer
- Morgantown
- Owensboro
- Catlettsburg
- Wilsonville
- Baytown

Legend:
- bituminous
- lignite
- sub-bituminous
- anthracite
- plants in operation
- plants planned or under construction

churia, Spain, South Africa, Scotland, Sweden and — during the 19th century —parts of the eastern United States. But with the discovery of extensive oil and natural gas fields in the 20th century, there was little incentive in this country to tap shale oil — save for a flurry of interest during World Wars I and II.

Gases derived from coal were used before the turn of the century both in the United States and in Europe. Until the 1920s, a low-grade coal-derived fuel called "town gas" was used in many parts of the country to fire street lamps and home ovens. When oil and natural gas became cheap and plentiful, town gas disappeared.

German, U.S. Synthetics in World War II

During World War II, Germany used synthetic oil to supplement its sparse petroleum supply. "For Germany, between 1919 and 1945, the question of oil — its production, importation, synthesization, stockpiling, allocation and consumption — occupied a status which was second only to the survival of the political state," wrote Arnold Krammer, associate professor of history at Texas A&M University.[14]

A scarcity of oil hurt Germany during World War I. After the war, the nation began to investigate how to convert its abundant

[14] Writing in *Technology and Culture,* July 1978, p. 394. The magazine is published by the University of Chicago Press.

reserves of lignite, anthracite and bituminous coal into gasoline. Research into the process of synthesizing ammonia led to the discovery in 1921 of a hydrogenation technique which converts coals, tars and other substances into high-grade liquid fuels, including motor oil and aviation fuel.

Other synfuel processes were discovered by German scientists during the 1930s, and when the Nazis came to power in 1933 the coal and chemical industries were given generous government support. By the outbreak of World War II in 1939, Germany had 14 synthetic fuel plants operating at full capacity. At peak capacity, the synfuels industry produced about 100,000 barrels a day.[15]

A big part of the U.S. rubber supply, from the East Indies, was cut off after December 1941 by the Japanese advance in Southeast Asia. Two years later, three American synthetic rubber plants using alcohol and benzene as raw materials were in operation, supplementing the continuing supply of raw rubber from Brazil. The number grew to 50 by the end of the war when annual production of synthetic rubber reached one million tons. A cooperative effort by government and industry developed the American synthetic rubber industry from scratch during the war. Is that effort comparable to the current movement toward synfuels?

"Lessons about cooperation [between government and industry] can be learned from the synthetic rubber program," said James D. D'Ianni, a Goodyear Tire and Rubber Co. researcher who coordinated synthetic rubber research during the war, "but it's not an identical situation by any means."[16] The main difference is that the crash rubber program was significantly smaller in scope than the proposed synfuels program. The government, for example, invested about $700 million in the synthetic rubber industry; the Carter plan for synfuels calls for $88 billion in federal funds. In addition, the nation had no choice but to produce synthetic rubber during the war. The need was pressing at that time; but today there are several options available other than synfuel development.

Alcohol's Availability as an Automotive Fuel

Biomass, the commercial production of alcohol from agricultural crops, residues and wood process wastes, has been categorized as a synthetic fuel. Biomass produces two types of fuel, methanol and ethanol alcohol, through fermentation and other processes. Ethanol can be blended with gasoline to produce

---

[15] The Germans used a number of synfuel processes to turn coal into liquid fuels. They were called Bergius-hydrogenation, Fischer-Tropsch, Linde-Karawat, Koppers, Lurgi-Drawe, Pintsch-Hillebrand, Bubiad-Didier, Winkler and Schmalfeldt.

[16] Quoted in *The Wall Street Journal*, July 24, 1979.

gasohol. Production of ethanol is more expensive than petroleum refining, but government incentives, including the proposed elimination of a 4-cent-a-gallon federal gasoline tax, could lower the price enough to make it competitive with gasoline.

The government estimates that about 60 million gallons of ethanol are being produced yearly in this country, and that ethanol is sold at 800 or more retail gasoline outlets. Energy analysts predict that five times as much will be produced by 1982 and ten times as much by 1985 — equaling about 5 percent of all the gasoline sold in the United States.

Alcohol as a carburetor fuel is not new. It was produced in large quantities in this country until oil became widely available and inexpensive. Henry Ford, the automotive pioneer, used farm-produced alcohol to power an automobile, the quadricycle, he developed in the late 1880s. The first Model A Fords were equipped with adjustable carburetors that could use either gasoline or alcohol.

The first large producer of ethanol was the Atchison Agrol Plant. "By 1938," Hal Bernton has written, "some 18 million gallons of grain alcohol (ethanol) were distributed to 2,000 independent service stations who proudly marketed the blends of 10 percent alcohol and 90 percent gasoline. The service stations owned by the major oil companies would rarely touch the stuff."[17] Agrol made alcohol from a mixture of barley, rye, corn grain, sorghum, Jerusalem artichokes and blackstrap molasses. Stiff competiton from the oil companies, though, soon forced Agrol to cut production drastically. There was a brief rejuvenation during World War II when Agrol returned to full production to help produce alcohol to make synthetic rubber, medicines and such war materiel as gunpowder, and aviation and submarine fuels. When the war ended in 1945, the government stopped supporting the alcohol industry and Agrol again was forced to cut back operations.

Brazil's Increasing Reliance on 'Gasohol'

Today's search for fuels has prompted new interest in biomass in the United States and abroad. The nation that has led in the production and development of alcohol for fuels is Brazil. Its large sugar industry has produced alcohol as a byproduct for half a century, and the government began promoting bigger production in 1973 when the price of imported oil began climbing steeply. Brazil imports about 80 percent of its oil.

Brazil's goal is to increase the amount of alcohol added to gasoline until eventually its autos burn nothing else. The na-

---

[17] Writing in *The Washington Post*, Aug. 5, 1979. Bernton, a reporter for columnist Jack Anderson, is writing a book on alcohol's use as a fuel.

tion's automobile manufacturers have begun building cars with engines that can run on gasoline with a high-alcohol content; today the average content is about 20 percent. In July, the Brazilian branch of Fiat started building a line of cars to run solely on ethyl alcohol. Other Brazilian manufacturers, including branches of Volkswagen and General Motors, plan to follow Fiat's example.

Brazil produced 686 million gallons of ethanol last year and the goal for 1980 is 1.3 billion. Brazilian energy experts say their country will need more than 200 new distilleries and some 3.7 million acres of new sugar cane plantings to reach the 1980 quota. Brazilian scientists are experimenting with making alcohol from other sources, including cassava plants and babacu, a tall palm tree that grows extensively in the northeastern part of the country.

Among the positive aspects of biomass production is the fact that the green plants used to make alcohol grow throughout the world, and nearly every nation could produce much or all of its own fuel. Many substances have been used to produce alcohol, including most grains, manure, cheese whey, citrus wastes, and municipal solid wastes. Farmers have set up their own biomass operations using ground corn, manure from dairy barns, spoiled hay or other agricultural waste products. Another advantage of biomass is that, unlike fossil fuels, the energy resources that make alcohol are renewable. Moreover, there are significantly fewer environmental problems associated with biomass production than with the refining of oil and coal.

There are some drawbacks. One involves the amount of energy consumed to produce the grain that goes into producing alcohol. When petroleum products are used in tractors and other machinery to cultivate, harvest, ferment and distill the crops used for the extraction of alcohol, sometimes more fuel is burned than is produced. In addition, there is a question about the efficiency of alcohol in the gas tank. Mileage test results of gasohol-powered vehicles have varied. Some tests indicated that gasohol gave slightly better mileage than straight gasoline, while others indicated slightly worse mileage.[18]

Milton Copulos, energy analyst for the Heritage Foundation, recently offered a conclusion that mirrors the opinions of many observers: "While keeping perspective and realizing that the use of alcohol fuels in and of themselves will not solve the energy crisis, we must acknowledge that they do present one of our most hopeful short-to-intermediate term solutions. . . ."[19]

---

[18] Methanol is used to power a type of vehicle that is not designed for fuel efficiency — racing cars such as those that run in the Indianapolis 500.

[19] "Alcohol Fuels: Energy from Agriculture," The Heritage Foundation, June 4, 1979, pp. 7-8.

Synthetic Fuels

The Carter administration has provided modest government incentives for wider production and use of alcohol as an engine fuel. These incentives include loan guarantees from the Department of Agriculture for alcohol pilot plants and the use of alcohol to power federal vehicles. A Department of Energy policy review report, issued July 11, predicted that in parts of the country rich in raw materials for producing alcohol, it will be used extensively as a fuel.

Nuclear Fusion: Hope for a Distant Future

While alcohol's contribution to the nation's fuel needs could lie in the immediate future, nuclear fusion is but a distant hope. Fusion, or thermonuclear power, is the reaction that produces hydrogen bomb explosions and is the energy source of the sun and other stars. But, unlike nuclear fission, fusion has never been harnessed for peaceful purposes.[20] Although most scientists agree that it could provide the world a source of clean and inexhaustible energy, estimates of the time needed for its development vary from 25 to 100 years. Fusion's radiation hazards would be negligible; its only waste product would be innocuous helium gas. Moreover, fusion carries no danger of a runaway reaction resulting in an accidental discharge of radioactivity.

Research on fusion power began secretly more than 25 years ago almost simultaneously in the United States, Britain and the Soviet Union. An international agreement in 1958 ended the secrecy and set a policy of cooperation which has endured. The European Economic Community, Japan and the Soviet Union have joined with the United States as equal partners in a program to develop and test a fusion reactor. At least one expert believes the United States is the leader in world fusion technology. Melvin B. Gottlieb, director of the Plasma Physics Laboratory in Princeton, N.J., told *Context* magazine that "despite the major and impressive efforts in the Soviet Union, Western Europe and Japan," the United States could develop a demonstration reactor before any other nation does — possibly by the year 2000.[21] The Department of Energy, which coordinates a $500 million research and development fusion program, estimates that such a reactor will not be developed until 2015.

Environmentalist Barry Commoner has written that the United States' current energy problem is so critical that the nation cannot afford to wait until fusion's complex engineering and physics problems are solved. "To embark, today, on a transition to a nonexistent but hoped-for renewable energy source" like fusion, Commoner wrote, "would be like building a

---

[20] For a discussion of nuclear fusion, see "New Energy Sources," *E.R.R.*, 1973 Vol. I, pp. 193-196.
[21] *Context*, Vol. 8, No. 2, 1979. p. 13. *Context* is published by E. I. du Pont de Nemours & Co.

bridge across a chasm without first locating the other side. . . . We can neither wait for the experiments to succeed nor take the risks involved in embarking, now, on a transition that depends on their future success."[22]

Potential Environmental Effects

THE FUTURE of synthetic fuels is closely linked to their potential effects on the environment. President Carter's synfuels proposals and the pending legislation in Congress have received low marks from environmentalists. Carter's energy policy "is a combination of the technologically spectacular and the environmentally disastrous," Janet Marinelli wrote in *Environmental Action* magazine, "a hard path solution based on nuclear power, untested synfuels development and unaccountable bureaucracy."[23] Many of those who fear the potential environmental harm of a synthetic fuels industry argue for mandatory conservation measures and for the development of non-polluting, easily renewable sources of energy such as solar power. The environmentalists' biggest concern is that Congress will allow current environmental protection laws to be bypassed in order to hasten the development of synthetic fuels.

"I do not pretend that all new replacement sources of energy will be environmentally innocuous," Carter said in an environmental message to Congress on Aug. 2. "I will work to ensure that environmental protections are built into the process of developing these technologies, and that when tradeoffs must be made, they will be made fairly, equitably and in the light of informed public scrutiny. . . . I will not allow it to undermine protection of our nation's environment."

Carter's message contained about a dozen environmental initiatives,[24] most of which were praised by environmental groups. There was negative reaction, though, to Carter's statement on synfuel plants and environmental protection laws. "It was a kind of tiptoe-through-the-tulips message," said Louise Dunlap of the Enviornmental Policy Center. "He studiously avoided any mention of the conflicts between energy development and environmental issues."

Concern Over High Carbon Dioxide Levels

One of the environmental issues directly related to the build-

---

[22] Barry Commoner, *The Politics of Energy* (1979), p. 49.

[23] *Environmental Action*, September 1979, p. 25.

[24] Including programs to aid soil conservation and protect endangered species, and plans to reduce urban noise and to designate 145 new national trails and 75 new hiking trails in U.S. national parks.

ing of a synthetic fuels industry is the fact that coal burning increases the amount of carbon dioxide in the atmosphere. Some scientists have been concerned for some time about a "greenhouse effect" on the Earth caused by excessive amounts of carbon dioxide in the atmosphere. Carbon dioxide permits sunlight to penetrate through it but prevents energy in the form of heat from radiating back into space. The global climate would get warmer if carbon dioxide builds up, and conceivably lead to a melting of the polar ice cap, flooding of the coastlines, changing of rainfall patterns and disruption of agriculture.

"Such a series of changes," a report to the Council on Environmental Quality prepared by a group of prominent environmental scientists said, "would have far-reaching implications for human welfare in an ever more crowded world, would threaten the stability of food supplies, and would present a further set of intractable problems to organized societies."[25] Scientists cannot agree whether the Earth has already entered upon a long-term cooling trend or a warming trend. Expert opinion is divided.[26]

There are a host of other potentially damaging environmental problems associated with synfuel plants. Most scientific authorities agree that the facilities currently envisioned to produce oil shale, and coal gasification and liquefaction will not meet current air pollution standards. "Emissions vary according to the technology," Mark Trautwein of the congressional Environmental Study Conference wrote recently, "but very large amounts of particulates, sulfur dioxides, nitrous oxides, hydrocarbons and other pollutants would be released, not only in end-use consumption, but also in mining and at the plant where the coal or shale is converted to synthetic petroleum."[27]

Wastes from Oil Shale and Coal Plants

All synthetic fuel plants would generate enormous amounts of solid waste. A recent Department of Energy report[28] cited "major uncertainties" involving the large amount of solid waste emitted in shale-oil processing. "Disposal of spent shale and storage of raw shale," the report said, "would create land disturbances of large magnitude, accumulation of toxic substances in vegetation and contamination of groundwaters and surface waters runoff." It is estimated that a 50,000-barrel-per-day oil shale plant would generate between 30,000 and 60,000 tons of spent shale a day by conventional recovery methods. A similar

---

[25] George Woodwell, et al., "The Carbon Dioxide Problem: Implications for Policy in the Management of Energy and Other Resources," report to the Council on Environmental Quality, July 1979, p. 1.

[26] See "Weather Forecasting," *E.R.R.*, 1979 Vol. II, pp. 85-104.

[27] Writing in *ECS Fact Sheet,* July 12, 1979, p. 5.

[28] "Environmental Analysis of Synthetic Fuel Plants," July 12, 1979.

coal liquefaction plant would give off 60,000 to 4 million tons of waste a year.

The synthetic fuel plants' extraordinarily heavy demand for water could cause serious problems in the arid West where water always has been precious.[29] Polluted discharges from synfuel plants are not likely to be absorbed into streams and lakes without raising pollution levels to unacceptable levels. And the large use of water could divert water from irrigating tens of thousands of acres of agricultural land.

Strip mining of coal adds to the problem of water pollution and soil erosion. Most of the coal needed for the proposed liquefaction plants would come from the West, where strip mining is already extensive and reclamation is made all the more difficult to achieve because of the dry climate.

Other environmental questions involve cancer-causing agents in coal liquefaction and shale oil plants, as well as in the use and transportation of the fuel. One known carcinogen, benzo-a-pyrene, has been found in processed shale oil. It is not known how much, if any, of it remains in a commercial plant's waste. "It is known that there are quantities that do exist on processed shale," said Terry Thoem, director of the Environmental Protection Agency's Energy Policy Coordination Office in Denver. "The question is, at what levels, what concentration?"[30] Moreover, research conducted by Dow Chemical Co. this year found evidence that dioxin, one of the most toxic substances known, is present in the emissions of coal-burning power plants.

The Department of Energy study of the environmental impact of liquid synfuel plants found that "occupational health is not now felt to be a constraining factor in development [of coal liquefaction plants], but many questions must be answered regarding worker exposure to process-associated materials." The study recommended the expansion and acceleration of occupational health studies "to answer as quickly as possible the questions raised by current knowledge of the materials involved in the process." The Environmental Protection Agency, the Department of Energy and the American Petroleum Institute currently are studying the extent of carcinogens in oil shale extraction and its wastes. Government analysts think it will take up to eight years to conduct a thorough study of the possible harmful effects of coal liquefaction plants alone.

Alternatives to a Synthetic Fuels Industry

Criticism of the synthetic fuels program has centered on high costs and environmental problems. There are also concerns

---

[29] See "Western Water: Coming Crisis," *E.R.R.*, 1977 Vol. I, pp. 21-40.
[30] Quoted in *The Washington Post*, Aug. 17, 1979.

about the fact that it will do very little to help the nation's current oil problem. What are the alternatives? A six-year study of the nation's energy situation by the Harvard Business School recommended that the United States use a balanced energy program that relies on the transition to solar energy rather than a massive synthetic fuel development program. The study's findings, released in book form in August under the title *Energy Future,* call for conservation measures such as government subsidies for more fuel-efficient automobiles and residential solar installations.

With such subsidies we "can save five million barrels of oil a day by the late 1980s — faster than you can get even one million barrels a day from synthetic fuels," said Robert Stobaugh, director of the Harvard Business School's Energy Project and co-editor of the book. "We're not saying we're going to have a solar society. We're just saying we should go in this direction, and the move should be supported not only by environmentalists but by the oil companies and the rest of the business community — for their own good."[31]

A coalition of five environmental groups — Friends of the Earth, The Wilderness Society, Environmental Policy Center, Sierra Club and the Natural Resources Defense Council — has proposed an energy course for the nation that does not depend on synfuels. In a letter to President Carter on July 12, the group wrote: "Conservation is our quickest, cheapest and most environmental energy option." The letter said that synthetic fuels will be too expensive, involve the use of too much untried technology, will do nothing between now and 1990 to ease the nation's dependence on foreign oil and pose too many environmental risks. Sens. Edward M. Kennedy, D-Mass., and John A. Durkin, D-N.H., plan to introduce legislation in Congress this fall to shift the emphasis toward conservation.

What role will synthetic fuels have in the nation's future energy picture? It is certain that in the next few years biomass, oil shale and coal liquefaction will provide significantly more of the nation's fuel than they do now. But it looks as if synfuels will not significantly ease the nation's dependence on foreign oil until 1990 at the very earliest. If Congress and the president do not agree on a synfuels program, the nation either will have to turn to conservation and the development of alternative energy sources like solar power, or face an extremely critical situation in the years ahead.

---

[31] Quoted in *The New York Times,* July 12, 1979.

Selected Bibliography

Books

Brown, Lester R., *The Twenty-Ninth Day,* Norton, 1978.
Commoner, Barry, *The Politics of Energy,* Knopf, 1979.
Hayes, Denis, *Rays of Hope: The Transition to a Post-Petroleum World,* Norton, 1977.
Rocks, Lawrence and Richard P. Runyon, *The Energy Crisis,* Crown, 1972.
Stobaugh, Robert and Daniel Yergin, eds., *Energy Future,* Random House, 1979.

Articles

"A Desperate Search for Synthetic Fuels," *Business Week,* July 30, 1979.
Bartlett, Bruce, "Synthetic Fuel Booty," *Inquiry,* Aug. 6, 1979.
"Coal Gasification for Electric Utilities," *EPRI Journal,* April 1979.
Davis, W. Jackson, "Energy: How Dwindling Supplies Will Change Our Lives," *The Futurist,* August 1979.
Flanigan, James, "The Methanol Age is Dawning," *Forbes,* Aug. 6, 1979.
Pelham, Ann, "Energy Conscious House Approves Synthetic Fuel Bill," *Congressional Quarterly Weekly Report,* June 30, 1979.
The Energy Daily, selected issues.
Thomsen, Dietrick E., "Reaching for the Stars," *Science News,* July 21, 1979.
U.S. Department of Energy, *Energy Insider,* selected issues.
Velocci, Tony, "Energy: Time to Get Moving," *Nation's Business,* August 1979.
Wade, Nicholas, "Synfuels in Haste, Repent at Leisure," *Science,* July 13, 1979.

Reports and Studies

Committee for Economic Development, "Helping Insure Our Energy Future: A Program for Developing Synthetic Fuel Plants Now," July 29, 1979.
Council for a Competitive Economy, "Synthetic Fuels: Economics and Politics," Aug. 9, 1979.
Editorial Research Reports, "Public Confidence and Energy," 1979 Vol. I, p. 383; "Oil Imports," 1978 Vol. II, p. 621; "America's Coal Economy," 1978 Vol. I, p. 281; "New Energy Sources," 1973 Vol. I, p. 187; "Oil Shale Development," 1968 Vol. II, p. 905.
Merrow, Edward W., et al., "A Review of Cost Estimation in New Technologies: Implications for Energy Process Plants," Rand Corporation, July 1979.
U.S. Department of Energy, "Environmental Analysis of Synthetic Fuel Plants," July 12, 1979.
——"Report of the Alcohol Fuels Policy Review," July 11, 1979.
U.S. General Acounting Office, "An Evaluation of Proposed Federal Assistance to Financing Commercialization of Emerging Energy Technologies," Aug. 24, 1976.
Woodwell, George, et al., "The Carbon Dioxide Problem: Implications for Policy in the Management of Energy and Other Resources," Report to the Council on Environmental Quality, July 1979.

DEEP-SEA MINING

by

Kennedy P. Maize

Oct. 6
1978

Editor's Note: The 11th round of talks in the United Nations Law of the Sea Conference got under way in New York on March 8, 1982. The Reagan administration has adopted a tough stance toward the conference. In March 1981 Reagan stunned the other nations participating in the conference when he announced, just six days before the 10th round of talks was scheduled to begin, that the United States was pulling out of the negotiations until the administration had a chance to review what had been agreed to so far. Until then, the 10th session was expected to be the concluding one — except for the formality of a later treaty signing.

After a 10-month review, the administration announced that the United States would return to the Law of the Sea Conference, but that the U.S. negotiators would seek to relax proposed curbs on deep seabed mining.

Even before the United States temporarily pulled out of the conference, there already were signs of trouble. Congress in 1980 incurred the wrath of many of the foreign delegations by passing a law — the Deep Seabed Hard Mineral Resources Act of 1980 — permitting U.S. mining companies to begin ocean-floor exploration in 1982 and actual mining in 1988. The companies argued that without such a national guarantee of access their bankers would not finance the ventures. The Third World delegates saw the law as a U.S. attempt to pre-empt the ocean's mineral resources and undercut the treaty-making.

Deep-Sea Mining

THE NATIONS of the world have been meeting twice a year since 1973 in a futile attempt so far to agree on a new international law of the sea. The latest gathering — the seventh session of the Third United Nations Law of the Sea Conference — adjourned Sept. 15 in New York, with little progress to report. Elliot L. Richardson, the chief U.S. negotiator at the conference,[1] said at adjournment that not much had been accomplished. But a "momentum toward a settlement" achieved at an earlier session in Geneva had been maintained, he added.

As in other recent sessions, issues over mining the ocean floor beyond the continental shelves *(see terminology, p. 165)* provided the chief barrier to a settlement. Third World nations favor an ocean mining "regime" run solely by a U.N. agency, while the United States favors an approach involving private enterprise. An added complication is the prospect that Congress, with the Carter administration's blessing, might pass legislation to permit seabed mining by U.S. companies before a U.N. treaty is negotiated *(see p. 170)*. Congressional action has been denounced by Third World spokesmen as disruptive and illegal. The U.S. delegation disputes that analysis.

Even before the New York meeting opened Aug. 22, Conference President A. Shirley Amerasinghe of Sri Lanka (Ceylon) told reporters that at least one more session would be needed to reach a final agreement. In adjourning the New York session, Amerasinghe announced that the conference would convene again in Geneva next March. Most observers doubt that the difficult issues can be resolved in just one more six-week negotiating session.

The current Law of the Sea Conference, the third U.N. effort since 1958 to rewrite fundamental international sea law, has resolved many of the issues that had sunk the first two conferences. The 158 nations currently represented at the conference have agreed in principle to limit their offshore sovereignty to 12 miles but maintain a zone of exclusive economic exploitation as far as 200 miles offshore. This is largely a recognition of conditions as they already exist. Most maritime nations currently

---

[1] Richardson, holding the rank of ambassador at large, is formally the President's Special Representative to the Law of the Sea Conference.

have 12-mile territorial limits and a 200-mile fishing or economic zone. There is also general agreement on free passage of ships through straits that now fall within the 12-mile territorial boundaries (many had been expanded from three miles).[2] And there has been progress toward agreement on what mechanisms to use for settling international maritime disputes. According to Alan Berlind, head of the State Department's Interagency Group for the Law of the Sea, the New York meeting resulted in agreement on pollution-control rules.

Territoriality, free passage and pollution control were the main factors behind U.S. support of a conference when it was proposed in 1967 *(see p. 174)*. For many of the Third World nations, on the other hand, ocean-mining was the issue that motivated them to push for a conference. Ocean mining has become the sticking point in all of the sessions since the second was held in Caracas, in 1974. At stake in the debate over ocean mining are important political and economic interests that pit the "have" and "have not" nations against each other. Charles Horner, senior legislative assistant to Sen. Daniel P. Moynihan, D-N.Y., called the conference "the most complicated diplomatic negotiation in all of history."[3]

U.S. Business Interests in Seabed Minerals

Many areas of the ocean floor are littered with mineral nodules — potato-shaped lumps of mud and minerals that include valuable amounts of manganese, copper, nickel and cobalt. The nodules typically consist of 24.2 percent manganese, 14 percent iron, 1 percent nickel, 0.5 percent copper and 0.35 percent cobalt. Though scientists are still studying their formation, the current hypothesis is that the nodules form around a "seed" of some sort — a shark's tooth or a piece of shell — and the minerals precipitate out of seawater and settle around the seed.

One area of the Pacific floor between Mexico and Hawaii, known as the Pacific Quadrangle, is particularly rich in nodules. Deposits of up to 100,000 tons per square mile are found there. The company or country that is able to mine those deposits will likely reap vast profits. *Nation's Business,* magazine of the U.S. Chamber of Commerce, has called the ocean floor "a treasure trove of immense proportion."[4] Jack Flipse, until recently head of Deepsea Ventures, a leading ocean-mining company, said his calculations in the late 1960s indicated that first-year profits from a deep-sea mining venture could run to $100 million. At that

---

[2] The U.S. Continues to have a 3-mile limit, but recognizes other states' 12-mile boundaries.

[3] Charles Horner, "Who Owns the Sea?" *Commentary,* August 1978, p. 61.

[4] Sterling G. Slappey, "Who Will Reap the Mineral Riches of the Deep?" *Nation's Business,* March 1978, p. 25.

Law of the Sea Terminology

Continental Shelf. As a legal term, the continental shelf is the seabed and subsoil of the submarine areas adjacent to the coast but outside the area of the territorial sea, extending to a depth of 200 meters (about 600 feet).

Deep seabed. The ocean floor beyond the outer limit of the continental margins. The ocean floor in the deep seabed is generally 12,000 feet and more below the surface. Also known as "the abyssal plain."

Economic zone. A 200-mile zone off the coast where the coastal nation has primary economic interest, and has the first opportunity to exploit the riches of the sea.

High seas. All water beyond the outer limit of the territorial seas.

Territorial sea. A zone off the coast where complete sovereignty is maintained by the coastal nation.

rate, his company could recover its total investment in just two years of operation.[5]

International business interests have taken the lead in developing the technology for mining the ocean floor. It is estimated that they have spent $2 billion preparing to harvest mineral nodules from as deep as 20,000 feet. According to the American Mining Congress,[6] eight international consortia are currently involved in developing the ocean mining industry:[7]

The Kennecott Group, including companies from the United States, Britain, Japan and Canada; Ocean Mining Associates, a partnership of three companies — two from the United States and another from Belgium; INCO Ltd. of Canada, consisting of Canadian, U.S., West German and Japanese companies; The Lockheed Group, including U.S. and Dutch companies; Afemod, a French consortium; DOMA of Japan; CLB Consortium, including Canadian, Australian, Japanese and U.S. companies; and Eurocean, made up of companies from France, Sweden and the Netherlands.

In the South Pacific last summer, the experimental mining ship SEDCO 445 recovered some 700 tons of nodules from a depth of 17,000 feet, causing the business publication *Barron's* to observe: "The test offered further proof of the fact that undersea mining, an idea worthy of Jules Verne, is in fact technically feasible."[8]

---

[5] See "Changing Profile of Deep-Sea Miners," *Science,* June 2, 1978. p. 1030.

[6] The American Mining Congress is the chief trade organization of the mining industry in the United States. Membership includes producers of coal, industrial and agricultural minerals and metals, manufacturers of mining equipment, and others with interests in mining.

[7] Data from a letter to Rep. John B. Breaux, D-La., of the House Merchant Marine and Fisheries Committee, from the American Mining Congress, May 26, 1977.

[8] *Barron's,* Sept. 4, 1978, p. 7.

The Lockheed Group recently refitted the 618-foot Glomar Explorer for an experimental ocean-mining voyage scheduled to begin early in November. The Glomar Explorer was built by Howard Hughes and used by the CIA in 1974 to recover portions of a Soviet submarine that had sunk in the Pacific. The Central Intelligence Agency disguised its activities by saying the ship was being used in ocean mining.

The business ventures interested in ocean mining have since 1970 adhered to a voluntary U.N. moratorium on commercial-scale ocean mining. Complying with that moratorium was done at no real cost to the companies, however, because their technology was not far enough advanced to permit them to begin large-scale operations. Most observers now feel that the companies will be ready to begin commercial operations sometime between 1985 and 1990. Marne A. Dubs of the Kennecott Copper Corp. has said that "deep-sea mining, by the early part of the 21st century, could provide more than half of the nickel and manganese the world consumes and all of its cobalt."[9]

Under traditional international law — the doctrine of freedom of the high seas — material on the ocean floor beyond territorial limits belongs to whatever party can get it. But that concept is being displaced by one set forth in 1967 by Arvid Pardo, Malta's ambassador to the U.N. Pardo said in a General Assembly speech that the resources of the sea should be "the common heritage of mankind" *(see p. 174)*. All of the 158 nations currently involved in the Law of the Sea Conference profess to accept Pardo's new doctrine. But applying the doctrine to mining has been impossible so far.

Politics of U.N. Law of the Sea Conference

The stalemate over deep-sea mining is deep-seated, a complex mixture of international politics, economics and ideology. Despite rhetoric about mankind's common heritage, matters of a more selfish nature guide the day-to-day bargaining. The matters concern strategically vital minerals, protection of existing industries, and the prospect of revenues amounting to billions of dollars. U.S. interests are both strategic and economic, with a touch of ideology mixed in as well. A recent study by the General Accounting Office, the investigative arm of Congress, concluded: "Deep ocean mining of manganese nodules by U.S.-based firms could benefit the national economy and may be important from a national security standpoint."[10]

The strategic value of ocean mining relates to U.S. dependence

---

[9] Quoted in *Barron's*, Sept. 4, 1978, p. 7.
[10] General Accounting Office, "Deep Ocean Mining — Actions Needed to Make it Happen," June 28, 1978, p. 43.

Technology of Deep-Sea Mining

Ten years ago, mining the ocean depths still awaited the right technology. The Marine Sciences Council predicted in 1968 that at least 20 years would pass before minerals other than oil, natural gas and sulfur could be extracted from shallow waters and even longer before being removed from the deep sea.

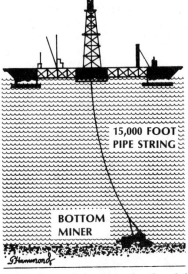

15,000 FOOT PIPE STRING

BOTTOM MINER

The technological barriers appeared impressive. Getting machinery to function on the ocean floor, under the great pressure of water overhead, would be difficult. Moving mineral nodules from the seabed to the surface would be an immense challenge, requiring innovations in equipment and in electronics to control the lifting devices.

But now, in 1978, three different approaches to mining the ocean floor appear to work. The technology for removing nodules from depths of three miles or more is considered to be proven. What remains involves engineering — scaling the demonstration models up in size for commercial usage. Thus the barriers to commercial exploration of the deep sea are political and economic, not technological.

A General Accounting Office Study of ocean mining said the three preferred mining systems are:

1. Suction dredge air lift, which works like a giant vacuum cleaner. The device sucks up the nodules and sends them to a surface ship by a pipe that trains from the ship. The vessel uses very complex and sensitive stabilizing devices to keep on a steady and even course.

2. Continuous bucket line, a Japanese-developed mechanism, picks up the nodules in containers somewhat like a waterwheel working in reverse. This device is less productive than the suction lift but better adapted to rugged deep sea terrain.

3. Hydraulic lift, a system pioneered by Lockheed and adapted from offshore oil drilling. An unmanned mining vehicle sweeps the ocean floor and pumps the nodules to a platform on the surface. This contrivance is apparently the most sophisticated of all; it has the greatest potential for production and for breakdown. Detailed workings of the Lockheed system are a closely guarded company secret. The illustration above shows its general outline.

on imported nickel, cobalt and manganese, each of which will be produced from the mineral nodules lying on the ocean floor. In recent years, according to GAO figures, the United States has imported 98 percent of its cobalt and manganese supplies, and 70 percent of its nickel. Manganese and nickel are both important in making steel. Cobalt is vital to heat-resistant materials used in turbines, jet engines, cutting tools and several other industrial and military products. A Stanford Research Institute study in 1974 cited in the GAO report ranked nickel, manganese and cobalt as among the minerals most critical to the U.S. economy and national security.

A fact that troubles U.S. strategic planners is that 70 per cent of the world's accessible cobalt (a byproduct of copper) is mined in Zaire's troubled Shaba province. The history of Zaire, once known as the Belgian Congo, is filled with political strife, civil war and disruptions by various forces that have tried to control the mineral riches in Shaba.[11] Ocean mining could move mineral production out of the troubled and politically volatile region. The question of offshore oil has not been a point of contention at the Law of the Sea Conference. Oil in geologic formations under the continental shelf will be covered by agreements on the 200-mile economic zone, and geologists say that finding oil under the deep seabed is not very likely.

The potential economic benefits to the United States from ocean mining are also significant. The GAO concluded that (1) the nation's imports of nickel, manganese, cobalt and copper could be reduced $1 billion or more by 1985; (2) jobs would be created by deep-sea mining, shipbuilding, the production of mining equipment, and the operation of mineral refineries within the United States; (3) exports of the mining technology and equipment to other nations would reduce the U.S. trade deficit; and (4) substantial exports of minerals would lead to a U.S. balance-of-payments surplus. These economic and strategic interests of the United States are shared by many industrial nations dependent on mineral imports and interested in the rapid development of ocean mineral sources. Japan and Western Europe generally support the United States in debates over ocean mining.

Opposing the industrial West have been a number of less-developed countries, allied into what has become known as the Group of 77. That name originated at the 1974 Caracas conference and is no longer accurately descriptive. Today the Group of 77 embraces as many as 120 Third World nations.[12] These nations have diverse economic interests in deep-sea mining but are united in their opposition to the industrial West.

---

[11] See "African Policy Reversal," *E.R.R.*, 1978 Vol. II, pp. 503-507.
[12] The current chairman of the Group of 77 is Satya N. Nandan of Fiji.

Deep-Sea Mining

Some of the countries in the Group of 77 are producers of land-based minerals that will be mined from the sea. These countries, such as Zaire, Zambia, Morocco and Gabon, often have little else of great economic value in their countries and depend on mineral revenues for their national wealth. They are unable to switch to other economic activities if mining profits dry up, and they lack the technological and economic capacity to engage in deep-sea mining. Consequently, they want to prevent rapid and wide-scale private exploitation of the minerals on the sea floor.

Several observers have suggested that powerful, developed countries which are also mineral exporters have been giving the Group of 77 behind-the-scenes support. The Soviet Union and Canada, both important mineral exporters, are often portrayed as silent partners with the 77 on ocean-mining issues. But Canada and the Soviet Union are playing a double game. Both countries are also pushing the development of ocean-mining technology — the Russians through government research and the Canadians through investment in various consortia.

Many nations in the Group of 77 are desperately poor, with no mineral reserves at all. They are pushing for an international mining system that will share its profits with all the nations of the world, regardless of their participation in mining efforts. These countries generally support the positions of the mineral-producing nations in the Group of 77. Some socialist countries, such as the United Republic of Cameroon, oppose the United States as a matter of ideology. Cuba is another country ideologically opposed to capitalism. But it is also an exporter of nickel — and that fact lends to ambiguity in its position with the Group of 77.

Ideology, it has been noted, is also a factor in the official U.S. attitude toward the conference. The American position has been that a free-enterprise approach to ocean mining is desirable and a socialistic solution to be avoided. Daniel S. Cheever, University of Pittsburgh professor of international affairs and a member of the international marine affairs panel of the National Academy of Science, said recently that "the possibility of a conference failure is due not only to the inflexibility of the Group of 77 but also to American ideological stiffness."[13]

Third World and U.S. Positions in Conflict

The industrial West initially pushed for the establishment of a U.N.-controlled agency to license businesses and national governments that wished to mine the sea. The agency would assign claims to various potential miners, collect a license fee, and

---

[13] Daniel S. Cheever, "The Law of the Sea: A Rejoinder to Richard G. Darman," *Foreign Affairs,* April 1978, p. 660.

and assess a royalty. The income from fees and royalties would be used to run the operation, with profits going to the U.N. The Group of 77, on the other hand, has pushed for direct control by a U.N. agency — an International Seabed Authority — with technology and financing provided by the industrial countries, and revenues distributed primarily to the underdeveloped nations of the world. The arm of the Seabed Authority that would actually mine the ocean would be called the Enterprise.[14]

What has evolved over the course of five years has been a compromise proposal for a dual system, providing both licensing of private groups and direct mining by the Enterprise. The technology would be provided by the business consortia as a condition for their mining. That compromise emerged in 1976 under the guidance of then-Secretary of State Henry A. Kissinger. The dual system seemed to provide a way out of the impasse that had developed. But the delegates have so far been unable to agree on some of the important details of the dual system. Richardson has described three key issues that remain to be resolved:

1. *The system of seabed exploitation.* The United States wants licensed companies and state-owned mining ventures to have unlimited access to mining sites. The Group of 77 prefers that the best sites be reserved to the Enterprise. A compromise will be difficult to achieve. "The issue," Richardson said, "is essentially one of economic pluralism versus state centralism projected on a global scale."[15]

2. *Policies on prices and production.* The issue represents a classic clash of interests between consuming nations and land-based producers. The United States, with its vast consumption, wants high production and low prices. Producer nations want just the opposite — high prices and low production. This issue, according to Richardson, "perhaps most sharply divides the conference."

3. *How the Seabed Authority would make decisions.* The key element is whether policies should be arrived at on a one-nation, one-vote basis, favoring the Group of 77, or on a basis that represents economic interests, rather than countries. The less numerous industrial countries fear the majority concept would lead to economic tyranny by the Third World.

Legislation in Congress to Permit Mining

Largely because of the U.N. stalemate, Congress has begun to act on behalf of American companies eager to mine the oceans. Legislation outlining rules for mining mineral nodules passed the House July 26 and moved to the Senate where it awaited floor

---

[14] See Elisabeth Mann Borgese, "Shaping the Law of the Sea," *World Issues,* October-November 1977, pp. 14-19.

[15] Speech to the Seapower Symposium of the Cincinnati Council of the Navy League of the United States, Jan. 18, 1978.

action as Congress pressed for mid-October adjournment. Legislation on ocean mining had been before Congress in previous years, but had never passed either house. Both the Nixon and Ford administrations opposed such legislation, fearing it would undercut the U.N. conference.

The Carter administration initially took the same line. But last January, it changed positions. In House testimony,[16] Richardson outlined the reasons for the shift. First, he said, the administration had concluded that some legislation would be necessary even if the conference were successful. The ratification process for any treaty the conference approved could take several years, and interim rules would be needed if companies were to mine during that period. And if no treaty would come out of the conference, Richardson added, legislation would provide the basis for international agreements between the United States and a number of other like-minded countries.

Many have seen the administration's changed position as an attempt to put pressure on the other countries at the conference. By encouraging companies to mine the sea, the United States would be signaling Third World countries to compromise or else watch events pass them by. However, State Department officials acknowledge privately that the United States, by acting alone to further its own interests, may strengthen the hand of "radicals" among the Third World delegates to the conference.

The bill that passed the House, HR 3350, would require companies and individuals to obtain permits from the Department of Commerce in order to mine the seabed beyond the continental shelves. They would also have to abide by the regulations intended to protect the environment. American companies would pay taxes on their profits from sea mining, even though the operations would geographically be outside U.S. jurisdiction. In addition, they would be required to turn three-fourths of one percent of their revenues over to an international fund to be shared with the other countries of the world.

Evolving Interest in the Sea

UNTIL THE mid-19th century, the study of the ocean depths was a minor preoccupation of a few seafarers interested in keeping their ships from running aground in shallow water. Yet the discoveries about the treasures that lie on the bottom of the sea stem from the careful and methodical application of the

---

[16] House Committee on International Relations, Jan. 23, 1978.

seafarers' ancient and simple technique of "heaving the lead" —
sounding the bottom of the ocean with line and lead weights.
The invention of the telegraph provided the stimulus to the
development of modern, accurate equipment for understanding
the sea's depths and the terrain of the ocean floor, leading to the
field now known as oceanography. In 1851 a telegraph cable was
successfully laid on the bed of the English Channel between
Dover and Calais. This prompted the idea of laying communica-
tions cables across the Atlantic, connecting the United States
and Europe. This was done in 1866.

This new activity also led to the outfitting of "the first major
effort of purely scientific exploration of the sea."[17] This was
the global voyage of the *H.M.S. Challenger* between 1872 and
1875. *Challenger* made soundings every 100 miles, including a
record sounding of 4,500 fathoms in the deepest of ocean
trenches, the Marianas Trench in the West Pacific. Scientists on
board the British vessel also devised equipment to bring up
samples of the ocean floor. *Challenger* dredged up a wide variety
of strange animals in the ocean depths, countering accepted
opinion that the temperatures and pressures of the deep would
not sustain life. Also dredged up were some potato-shaped lumps
of mud and minerals.

These mineral nodules were primarily objects of scientific
curiosity until after World War II, when new photography
techniques revealed vast concentrations of them on the sea floor.
By the late 1950s growing industrial demand for minerals raised
the possibility that the deposits might be commercially valuable.
Exploratory and scientific analysis began in earnest.[18]

Evolution of Freedom of the Seas Concept

Mineral nodules — seemingly a minor discovery by *Challenger*
— eventually provided a challenge to traditional concepts of the
international law of the sea. This law had evolved through centuries by custom and treaty as a mechanism for resolving disputes over fishing, navigation and trade.[19] At the base of the con-
ceptual framework of the law of the sea was the fact that the
oceans could not be permanently secured or occupied — fishing
boats and other vessels were only temporary occupants of the
vast ocean surface.

The theories of Hugo Grotius, the eminent Dutch scholar and
jurist of the early 17th century, became the foundation of the law
of the sea that most nations have accepted to this day. In *Mare
Liberum* (Free Sea), published in 1609 to defend the right of the

---

[17] Brenda Horsfield and Peter Bennet Stone, *The Great Ocean Business* (1972) p. 43.

[18] See Seyom Brown, et al., *Regimes for the Ocean, Outer Space, and Weather (1977)*, p.
73.

[19] See "Oceanic Law," *E.R.R.*, 1974 Vol. I, pp. 405-426.

Dutch or any other traders to operate in the East Indies, Grotius asserted that property ownership is based on possession or occupation. Things which cannot be seized or surrounded cannot be considered property, but are common to all. He argued:

> The air belongs to this class of things for two reasons. First, it is not susceptible of occupation and second, its common use is destined for all men.... Since the sea is just as insusceptible of physical appropriation as the air, it cannot be attached to the possessions of any nation.

The Grotian doctrine of freedom of the high seas became the basis of international sea law. Freedom of the high seas meant that no country could claim sovereignty or jurisdiction over the seas and all were free to compete for its riches as best they could. There was, fundamentally, no government or regime to control access to the seas.

Countries continued to claim sovereignty over narrow strips of the sea along their coastlines for self-defense and the protection of coastal commerce. Early authorities were unable to agree on an exact and uniform distance for these territorial waters, but a generally accepted distance of three miles was common by the early 20th century. But the three-mile limit was by no means universal and as new strategic concepts took form after World War II, pressure began building for revisions in traditional concepts related to the law of the sea.

New Challenges to Traditional Oceanic Law

President Harry S Truman on Sept. 28, 1945, declared that the United States would regard the natural resources of the continental shelf contiguous to its shores as subject to U.S. jurisdiction and control, regardless of the three-mile limit. A month later the president of Mexico made a similar declaration. Over the next dozen years numerous maritime nations followed suit. By 1958 pressure for a new look at international sea law had become irresistible.

The first United Nations Law of the Sea Conference met in Geneva in 1958 and approved four conventions — treaties — on the continental shelf, the territorial sea, the high seas, and fishing. But the conference was more notable for what it did not say, or said unclearly, than for what it defined and determined. The breadth of the territorial seas was left unspecified. The continental shelf was so vaguely defined that it seemed to extend as far as a nation wished to take it.[20]

---

[20] The conference defined the continental shelves as submerged coastal lands extending to a depth of 200 meters (about 600 feet) "or beyond that limit to where the depth of superadjacent waters admit of the exploitation of the natural resources of the...sea." The convention recognized that maritime nations had the exclusive right to explore and exploit the shelf adjacent to their shores and its natural resources.

The second U.N. Law of the Sea Conference assembled in Geneva in 1960. This conference again failed to define the limits of the territorial sea. A joint U.S.-Canadian proposal to allow states to claim a six-mile territorial limit, with a 12-mile fishing zone, failed by one vote to get the necessary two-thirds for adoption at the 1960 conference.

Sea as the 'Common Heritage of Mankind'

By 1967 there was again momentum for a new try at revising old concepts to fit new maritime realities. Less-developed countries had a growing fear that the industrial nations would appropriate the mineral riches of the ocean for their own use. Such action would be completely legal under the concept of freedom of the high seas — which makes the seas available to all who can exploit them.

At a meeting of the World Peace Through Law Conference in Geneva in July 1967, more than 2,000 lawyers urged the United Nations General Assembly to assume jurisdiction over the natural resources on the ocean floor. In August 1967, Malta's U.N. Ambassador Arvid Pardo offered a resolution in the General Assembly calling for a treaty to give the U.N. title to the sea-bed beyond territorial waters, establish an international agency to administer the resources in the sea-bed, allocate the revenues from the seabed to less developed countries, and ban weaponry on the sea floor.

In addressing the General Assembly the following November, Pardo asked for recognition of the principle that the ocean floor is "the common heritage of mankind." This would require a new approach to traditional sea law concepts. "Current international law encourages the appropriation of the ocean floor by those who have the technical competence to exploit it," Pardo said.

For the United States, the interest in changing the law of the sea in 1967 was primarily a matter of stemming the tide of "creeping jurisdiction" outward from national shorelines.[21] In the decade following the close of the first Law of the Sea Conference, an expansionist pattern of claims had developed. While 54 percent of the coastal nations claimed a three-mile territorial limit in 1958, 10 years later the number had dropped to 35 percent. In the same period, the number claiming territorial seas of 12 miles or more had increased from 18 to 43 percent.

For naval powers, including the United States, this "creeping jurisdiction" posed potential security dangers and a hindrance to commerce. A universal 12-mile territorial sea would overlap 121 international straits and thus could subject ships to tolls and

---

[21] So described by Richard G. Darman, "The Law of the Sea: Rethinking U.S. Interests," *Foreign Affairs,* January 1978, p. 375.

other restrictions imposed by the coastal nations. In the case of warships, permission for passage might be required. A different kind of objection to the 12-mile limit has been voiced by Dean Rusk, the former Secretary of State who now teaches at the University of Georgia School of Law. He forsees a conflict between the federal government and the coastal states over who controls the additional nine miles of territorial seas. Under the 1953 Submerged Lands Act, coastal state boundaries extend to the three-mile limit and these states have title to the lands beneath those waters.[22]

In 1970 the U.N. General Assembly voted to convene a third conference in 1973 to deal with the seabed, the continental shelf, fishing and conservation, the preservation of the marine environment and scientific research. Pardo again entered the scene when, in 1971, he set forth his concept of a new legal order for the oceans as a whole. This subsequently became articulated in Pardo's "Preliminary Draft Treaty for Ocean Space," which Elisabeth Mann Borgese of the International Oceans Institute has called the "prototype for the Single Negotiating Text," the document around which most of the negotiations have occurred.[23]

Results of Third Law of Sea Conference

After an organizational session in New York in 1973, the Third Law of the Sea Conference got down to hard negotiating in Caracas in the summer of 1974. After creating three substantive committees,[24] the delegates began a general discussion of basic issues. Borgese has described the Caracas session as dominated by "revolutionary ferment and the ascent of the Third World countries."[25] The third session, between March and May of 1975 in Geneva, resulted in the development of the informal Single Negotiating Text which became the basis for the discussions to follow. That text generally reflects the continued dominance of the Third World countries over the conference.

A fourth session followed a year later in New York. A revised Single Negotiating Text emerged from the New York meeting. But the conference began to become polarized and deadlocked over issues related to ocean mining. That deadlock continued through the fifth session, held in August and September 1976 in

---

[22] See Milner S. Ball, "The Law of the Sea, Federal-State Relations and the Extension of the Territorial Sea," a monograph prepared for the Dean Rusk Center, University of Georgia School of Law, Aug. 24, 1978. See also "Global Waterways: Access and Control," *E.R.R.*, 1967 Vol. II, pp. 832-837.

[23] Elisabeth Mann Borgese, "A Ten-Year Struggle for Law of the Sea," *Center Magazine*, May-June 1977, p. 53.

[24] Committee I handles ocean mining; Committee II concerns territorial seas, straits and economic zones; Committee III deals with marine scientific research and environmental protection.

[25] Borgese, *op. cit.*, p. 54.

New York. Only the appearance of Secretary of State Henry A. Kissinger at the conference and his suggestion for a dual public-private mining "regime" gave any hope for progress.

The sixth session, held in New York in the summer of 1977, appeared to be somewhat successful, reaching substantial agreements as to territorial seas, economic zones and the high seas. But the session fell apart toward the end of July, running aground on the rocky issue of seabed mining. The drafting had been left in the hands of the committee chairmen. Paul Engo of the United Republic of Cameroon, chairman of Committee I, made several major changes in compromise language that had been carefully negotiated by Jens Evensen of Norway. This turn of events led Elliot L. Richardson to declare that Engo's text was "fundamentally unacceptable."

It was at this point that Richardson and the Carter administration reevaluated their opposition to congressional legislation. In assessing the course of the conference prior to resuming the seventh session in New York in August, Richardson took a tough stance. "The forces now in motion," he said, "cannot be blocked or turned back. The seabeds contain minerals important to economic growth. They will be mined sooner or later, whether or not there is a treaty."[26]

Coming Issues in Sea Mining

EVEN IF the nations agree on the difficult issues that face the law of the sea conference, another hurdle remains. An international treaty requires ratification by the democratic countries involved, a prospect that may be very difficult, particularly in the United States. Without U.S. approval, the treaty would be worthless.

If a final agreement substantially resembles what has been agreed to so far, most analysts suggest that the American business community will oppose the treaty strongly when it comes before the Senate for ratification. *Nation's Business* concluded in March: "It is not in the public interest at this stage to delegate to any international agency the rights to marine resources. The existing national and international framework is considered sufficiently flexible to assure orderly development of the natural resources of the sea."

There is substantial political support for this position. Recent-

---

[26] Quoted by the Associated Press, Aug. 21, 1978.

New means of catching and processing fish are rapidly out-stripping the capacity of the sea's creatures to replenish themselves. Between 1951 and 1971 the global fish catch quadrupled. Fish now account for 10 percent of the world's pro-tein. As a consequence, scientists report drastic declines in catches of certain species of herring, cod, sardines and salmon.

The possibility that the bounty of the seas could vanish has led to a basic agreement at the Law of the Sea Conference. The agreement, embodied in an "informal composite negotiating text," requires that fishing within the 200-mile economic zone must be managed according to a scientific determination of the "maximum sustainable yield." Coastal states have overall jurisdiction over the zone and preferential harvesting rights in it. The United States in 1976 was the first major country to establish a 200-mile economic zone, an action many credit with saving the economic life of the Atlantic coast fishing fleet.

Most experts agree that the "maximum sustainable yield" concept works well for coastal fish species such as cod, flounder and halibut, which tend to remain in one area throughout their lifespan. For anadromous species — those that spawn in rivers and swim out to the high seas during their lifetimes, such as salmon — different rules are required. The conferees have agreed that taking anadromous species on the high seas would be banned except in special circumstances.

Highly migratory species, such as tuna, pose a different problem. These species cover vast distances through the waters off many nations, moving both within and beyond the economic zones. The conference has agreed to regional fisheries organiza-tions to fix conservation rules and fishing rights for these species.

In dealing with controversial fishing issues, the conference can claim considerable success. Even if a formal treaty is never forthcoming, many expect the agreement on fishing to remain in-tact. The conference has already been credited with ending the "cod war" between Britain and Iceland and with cooling disputes between the United States and countries such as Peru and Ecuador over tuna fishing.

ly Sens. Henry M. Jackson, D-Wash., and Clifford P. Hansen, R-Wyo., sent a letter to Richardson urging a harder line in the conference. They suggested "a careful examination of whether or not the United States Senate is prepared to ratify a law of the sea treaty...which would inhibit and prevent deep seabed development by U.S. firms." They added: "We are inclined to believe the Senate would not."[27]

Opponents of the treaty are considered likely to rely on the arguments outlined by Richard Darman in his influential article in *Foreign Affairs*. Darman argued that U.S. interests were no longer at stake in the conference. Indeed, he suggested, long-range U.S. interests may be best served by no treaty at all. Darman suggested that the United States "should exhibit a clear-eyed willingness to accept conference failure as a non-disastrous, indeed, thoroughly tolerable outcome." Some State Department officials have privately suggested that this is the direction U.S. policy has followed in recent months; they see the reversal of position on legislation as the starting point.

Environmental Problems to be Overcome

Before deep-sea mining by U.S. companies can begin, important environmental questions must be answered. Both the Law of the Sea Convention on the High Seas Treaty of 1962 and the National Environmental Policy Act of 1969 require an environmental assessment before mining can commence. Ocean mining poses several environmental threats. Destruction of sea-floor organisms and habitats could result from mining. Clouds of sediment stirred up by mining devices could cause pollution at the sea floor. Lifting the nodules to the surface could pollute upper levels of the sea with sediment. Finally, processing and refining nodules, either at sea or in plants onshore, could pose environmental hazards to the fragile coastal zone.[28]

Two Department of Commerce studies of the environmental impact of ocean mining are scheduled for completion in 1979. One is by the Office of Marine Minerals, dealing with the onshore processing of nodules. The other is the Deep Ocean Mining Environmental Study (DOMES), conducted by the department's National Oceanographic and Atmospheric Administration. The two studies will make up the environmental impact statement required by the law before mining can begin.

DOMES, a study conducted in two phases, is also intended to help the mining industry in designing mining equipment and operational techniques so as to have as small an impact on the marine environment as possible. Phase I, which is completed, es-

---

[27] Quoted in *Barrons*, Sept. 4, 1978.

[28] See General Accounting Office report "Deep Ocean Mining — Actions Needed to Make it Happen," June 28, 1978, pp. 18-19.

tablished preliminary environmental guidelines for mining the mineral nodules. Phase II, which is under way, involves actual monitoring of industrial mining tests. The objective of this phase is to check the accuracy of the material developed in Phase I, leading to the issuing of regulations for ocean mining. Those regulations, according to NOAA officials, are several years away.

Economic Factor in World Mineral Supply

Economic events may ultimately have more to do with the timetable for deep-sea mining than the political issues now facing world councils. In recent months, ocean mining companies have been scaling down their programs and setting back schedules. Where many company spokesmen had been predicting mining of the ocean floor by 1985, most now look to 1990 or beyond. John Shaw, president of Ocean Management Inc., was among those who earlier predicted a 1985 start. He now guesses that "the first commercial system certainly will be in the late 1980s or early 1990s."[29]

The problem is economic. The price of nickel, which many experts believe is the key element in profitable operations of an ocean mining venture, is depressed by overproduction. Market analysts see little chance that the price will rebound. Nickel is now selling for between $2 and $2.10 a pound. To justify the vast investment needed for an ocean mining venture, according to Marne A. Dubs of Kennecott, nickel prices would have to be in the range of $2.80 to $3.20 a pound. However, political events can sometimes quickly change expectations about world mineral supplies.[30] Cobalt prices, for example, have increased fivefold in nine months as a result of disruptions to the copper mines in Zaire's Shaba province.[31] The disruptions came from an invasion by rebels based in neighboring Angola.

Part of the recent pessimism of ocean mining companies is related to developments at the Law of the Sea Conference. Business interests fear that the Third World will reserve the best mining sites for the Enterprise, taking all the profit out of mining. If that occurs, the companies say, they will not enter the business. Ocean mining is clearly an issue with important ramifications far beyond the technological and engineering issues involved in converting mineral nodules into commercial metals. Matters of the shape of international institutions, ideological concerns, strategic considerations, and feelings of national pride have all become important to the question of who will reap the riches of the deep. In those circumstances, a quick and clear-cut solution is not likely.

---

[29] Quoted in *The Wall Street Journal,* July 31, 1978.
[30] See "World Mineral Supplies," *E.R.R.,* 1976 Vol. I, pp. 383-402.
[31] See "Industry's Scramble for Cobalt Supplies," *Business Week.* August 28, 1978, p. 40.

Selected Bibliography

Books

Bardach, John, *Harvest of the Sea,* Harper & Row, 1968.

Brown, Seyom et al., *Regimes for the Ocean, Outer Space, and Weather,* The Brookings Institution, 1977.

Horsfield, Brenda and Peter Bennet Stone, *The Great Ocean Business,* Coward McCann & Geoghegan Inc., 1972.

Articles

Borgese, Elisabeth Mann, "A Ten-Year Struggle for Law of the Sea, *The Center Magazine,* May-June 1977.

—— "Shaping the Law of the Sea," *World Issues,* October-November 1977.

"Changing Profile of Deep-Sea Miners," *Science,* June 2, 1978.

Darman, Richard G., "The Law of the Sea: Rethinking U.S. Interests," *Foreign Affairs,* January 1978.

Horner, Charles, "Who Owns the Sea?" *Commentary,* August 1978.

"Industry's Scramble for Cobalt Supplies," *Business Week,* Aug. 28, 1978.

Johnson, Kathleen S., "Law of the Sea Conference Goes On," *European Community,* September-October 1978.

Pelham, Ann, "Investment Guarantees Are Key Issue in Proposals to Promote Seabed Mining," *Congressional Quarterly Weekly Report,* Jan 21, 1978.

"Shark and Prey," *Barron's,* Sept. 4, 1978.

Slappey, Sterling G., "Who Will Reap the Mineral Riches of the Deep?" *Nation's Business,* March 1978.

Reports and Studies

Editorial Research Reports, "Oceanic Law," 1974 Vol. I, p. 403; "Oceans and Man," 1968 Vol. I, p. 320; "World Mineral Supplies," 1976 Vol. I, p. 383.

Department of State, "A Constitution for the Sea," August 1976 (Publication 8870, International Organizations and Conferences Series 123).

General Accounting Office, "Deep Ocean Mining — Actions Needed to Make It Happen," June 28, 1978; "Need for Improving Management of U.S. Oceanographic Assets," June 16, 1978; "Results of the Third Law of the Sea Conference 1974 to 1976," June 3, 1977.

AMERICA'S COAL ECONOMY

by

Kennedy P. Maize

Apr. 21
1 9 7 8

Editor's Note: On March 27, 1981, exactly three years to the day after the 1977-78 coal strike ended, 160,000 miners in the Eastern coal fields walked off their jobs after their old contract expired. The miners finally began returning to work on June 8, 1981, after ratifying a new 40-month contract negotiated by the United Mine Workers.

As a result of the 72-day strike, coal production in 1981 was down slightly from 1980. But consumption of coal was up slightly and exports increased. Utilities continue to use more coal, generating over half of the nation's electricity from coal.

The U.S. Supreme Court on June 15, 1981 unanimously upheld the Surface Mining Control and Reclamation Act of 1977, which gave the federal government broad powers to curb the abuses associated with strip mining. Federal judges in Virginia and Indiana had struck down provisions of the law that required companies to restore the land after mining.

Supporters of coal slurry pipelines *(see p. 189)* have organized a high-powered lobbying coalition — the Alliance for Coal and Competitive Transportation — to help them win their 20-year legislative struggle in Congress. The Focus of their campaign is on legislation that would give pipeline companies, under certain conditions, federal eminent domain power to secure easements across private land, including properties owned by the railroads. Eminent domain is the authority to take property for fair compensation.

AMERICA'S COAL ECONOMY

THE NATION has returned to normal after the disruptions of the recent coal strike. But as the memory of the bitter, 110-day strike fades, serious doubts remain about coal's role in meeting the nation's energy needs. President Carter's National Energy Plan, now stalled in Congress, calls for greater reliance on coal to ease America's dependence on foreign oil. But many observers fear that Carter's goal of almost doubling coal production and consumption by 1985 is unrealistic, especially in light of the strike. "The problem of labor stability [in the mines] is causing some serious skepticism about added dependence on coal," observed journalist Walter S. Mossberg shortly before the strike ended on March 27.[1]

Even before miners walked off the job last December there were doubts about Carter's plan to double coal output to 1.2 billion tons by 1985. Many experts question whether the coal industry can find enough money, machinery and miners to support such a rapid increase in production. Analysts also doubt that the nation's transportation system — especially the troubled railroads — can haul the coal even if it is produced. Environmentalists worry that increased coal production will result in more air and water pollution problems.

Despite the potential problems, the administration remains committed to a coal-based energy strategy. The reason is obvious: coal's sheer abundance. According to the National Coal Association, coal represents 80 per cent of America's total energy reserves; oil and gas make up only about 8 per cent. The U.S. Geological Survey says there are at least 1.7 trillion tons of coal beneath American soil. Estimates vary as to how much coal is recoverable by current technology. The Interior Department's Bureau of Mines estimates that the "demonstrated reserve base" — coal deposits at depths similar to those now being mined — is 438 billion tons. "Recoverable reserves" — that portion of the demonstrated reserve base that can actually be mined under current technological, economic and legal constraints — are put at about 219 billion tons.

The General Accounting Office is skeptical of the Bureau of Mines' estimates. It reported that the figures are based on a

[1] *The Wall Street Journal*, March 16, 1978.

"lack of consistent and reliable data and a lack of analysis of economic, technological and legal conditions on a site-specific basis to determine which reserves may or may not be mined." The GAO report did not specify the size of error in the Bureau of Mines' estimates, saying only that they are "extremely spotty and outdated."[2]

Coal's Role in Carter's Energy Program

The purpose of President Carter's National Energy Plan is to reduce American dependence on oil and gas. The administration has said the entire plan would save the nation 4.5 million barrels of oil a day; more than half of the total savings — 3.3 million barrels — would come from coal conversion alone. In his energy message to Congress on April 20, 1977, Carter said. "We must be sure that oil and natural gas are not wasted by industries and utilities that could use coal instead."

To encourage industries and utilities to switch from oil or gas to coal, Carter proposed a complex package of tax and regulatory measures. Under Carter's plan:

● No new industry or utility boiler would be allowed to burn oil or gas, with limited exceptions for extreme environmental or economic circumstances.

● All new facilities, even those burning low-sulfur coal, would be required to install the best technology available to control pollution.

● Natural gas would be banned as a fuel for electric utilities by 1990 and for other industries by an unspecified date.

● Existing facilities could be forced to convert to coal under some circumstances.

● Oil and gas burned by utilities and industries would be subjected to new taxes to make coal even more competitive as a boiler fuel.

● Tax rebates and credits to industries and utilities would be used to subsidize conversion from oil or gas to coal.[3]

The House passed a coal conversion bill on Aug. 5, 1977, but it was considerably weaker than the administration proposal. The White House wanted to impose heavy taxes on large gas and oil users to encourage them to convert to coal. The House bill provided so many exemptions to the taxes that the administration estimated it would reduce total energy savings by more than half. The Senate coal conversion bill, passed Sept. 8, 1977, further weakened the administration's plan

---

[2] GAO Letter Report EMD-78-23, "Summary of Problems Associated with Coal Reserve Estimates," Jan. 11, 1978.

[3] See *Congressional Quarterly Weekly Report,* June 18, 1977, pp. 1211-1218. *Report,* June 18, 1977, pp. 1211-1218.

National Energy Plan — Phase Two

The Carter administration has said it will announce in May a second phase of the National Energy Plan which the President unveiled in April 1977. Several elements of the plan already have been reported in the press, through news leaks presumably intended to prepare Congress and the public for what is to come. The news stories indicate that the emphasis will be on increased development of new and unconventional energy sources, and a shift away from nuclear energy. The second phase focuses on energy and supplies for the 1990s, while the initial plan concentrated on the period up to 1985.

R&D — research and development — is the key to the plan, especially in regard to coal. The administration envisions the use of federal funds to aid the commercial development of coal liquefaction and large-scale coal gasification plants, and tax credits or grants to finance new methods of cleaning up the burning of coal.

Even before it has fully emerged, the plan has drawn criticism. Environmentalists say it may put too much reliance on coal and not enough on solar energy.

In a report on the Senate bill issued July 25, 1977, the Senate Committee on Energy and Natural Resources noted that it included provisions that would reduce oil and gas savings by at least 200,000 barrels of oil per day in 1985. One of the bill's provisions greatly expanded the number of new industrial plants that would be allowed to burn oil — but not natural gas — instead of coal. Coal conversion legislation has been languishing in the House-Senate energy conference committee since last fall. Most observers expect the conference committee to act on coal conversion sometime this spring.

Labor Relations and Productivity Declines

Controversy has surrounded Carter's coal conversion plan from the start. Several preliminary analyses suggested that the administration's goals for increasing coal production were impractical. The General Accounting Office, in a report issued Sept. 22, 1977, concluded that "so many elements would have to work to double coal production by 1985 that GAO does not believe it could happen."[4] The GAO report listed several problems it felt could delay beyond 1985 the achievement of production levels the administration is seeking. The list included "long lead times required to open mines, environmental constraints, time problems in delivery of heavy equipment, capital problems, and labor and productivity problems." Miner productivity has been falling steadily since 1969, to the point

[4] General Accounting Office, "U.S. Coal Development — Promises, Uncertainties," Sept. 22, 1977, p. xv.

where today's miners are about as productive as those of 1960
(see p. 187).

Coal experts have identified several factors that account for
the productivity decline. Among them are wildcat strikes,
absenteeism and the introduction of large numbers of new, untrained miners. An important factor in the productivity drop is
said to be the Coal Mine Health and Safety Act of 1969 which
increased the number of people in the mines performing non-
mining, safety-related jobs.[5]

Deputy Energy Secretary John O'Leary acknowledged the
productivity problem in a recent interview.[6] He said that "a
productivity increase of just one ton would wipe out the inflationary aspects of the pay increase in the new coal contract."[7]
O'Leary said he expects to see productivity figures climb in the
near future because younger miners would have become more
adept at their jobs. He predicted that increased productivity
might make a real contribution to labor stability in the coal
fields. "It will build an argument for profit sharing," he said,
"which ought to help improve labor relations significantly."

Labor relations could be the key to meeting Carter's coal
production goals. Department of Labor figures on coal industry
strikes show an increased number of work stoppages, particularly wildcat strikes, in the past few years. These strikes trouble industry greatly and were one of the main issues in the
negotiations over the current contract. Mine operators wanted
the right to fire leaders of wildcat strikes and to have the union
fined for days of production lost due to such strikes. But the
final coal contract eliminated any such provisions, and wildcat
strikes may well remain common in the coal fields.

According to the General Accounting Office, the 2 per cent of
total working time lost each year because of wildcat strikes "is
not substantial."[8] Big losses in productivity do occur in years
when a contract between the United Mine Workers and the coal
industry expires. In 1974, for example, when the contract that
expired last December was negotiated, 8 per cent of total work
time was lost due to strikes.[9] Although the calculations have not
been completed for the period including the strike just ended,
they will certainly show a great loss in working time.

The labor-management climate in the mines also worries the

---

[5] Coal fatalities have declined substantially since 1970, from a rate of 1.02 per million
worker-hours to .35, according to the Bureau of Mines. Disabling injuries have dropped from
a rate of 45.40 per million worker-hours to 36.16.

[6] With Editorial Research Reports, March 28, 1978.

[7] The new contract, signed March 25, 1978, provides a 38 per cent pay increase over a
three-year period.

[8] "U.S. Coal Development — Promises, Uncertainties," *op. cit.*, p. 4.8.

[9] U.S. Bureau of Labor Statistics, "Collective Bargaining Summary, 1974," p. 5.

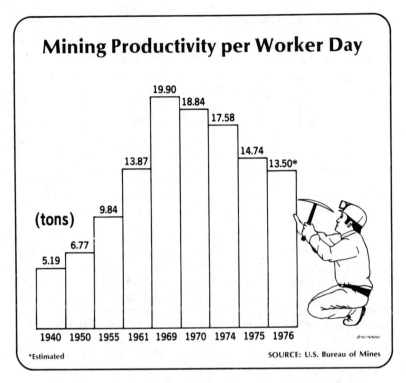

Mining Productivity per Worker Day

19.90　18.84　17.58　14.74　13.50*　13.87　9.84　6.77　5.19

(tons)

1940　1950　1955　1961　1969　1970　1974　1975　1976

*Estimated

SOURCE: U.S. Bureau of Mines

Carter administration. Energy Secretary James R. Schlesinger recently said: "The coal strike...has taught the nation the absolute necessity of achieving long-term stability in the mines. The industry is just going to have to be induced to institute proper labor relations. Stability will come only from a changed spirit in the coal fields."[10] President Carter announced March 25 that he would appoint a commission to investigate long-range problems in the coal industry and that labor relations would be given a high priority.

Despite warnings that coal production cannot be doubled by 1985, both the government and industry remain optimistic. Deputy Energy Secretary O'Leary said, "It's a hard goal, and it's not made easier by the fact of the coal strike. But it's achievable." Industry spokesmen take about the same line, though they couple their optimism with a request for less government regulation of their business.

In a letter to President Carter dated Jan. 12, 1978, Carl Bagge, president of the National Coal Association, said that the "goal of doubling coal production and use by 1985 will be achieved if utility and industrial organizations are allowed by the government to implement their current plans." Bagge pointed out that coal production in the September-November

[10] Quoted in *The Wall Street Journal,* March 16, 1978.

187

period prior to the strike was running between 15 and 16 million tons a week, demonstrating that "the nation already has capacity in place to produce 775 to 800 million tons per year."

A recent article in *Fortune* magazine argued that the nation is moving toward a surplus of coal, and predicted that "the belligerent and undisciplined U.M.W. will control a smaller share of the nation's coal output" as time passes, diminishing the threat of strikes and production disruptions.[11] If these assessments are accurate, there should be no problem of doubling coal production by 1985. But two problems will remain: how to move the coal to where it's needed and how to burn it without causing serious environmental hazards.

Bottlenecks in U.S. Transportation System

About two-thirds of all coal shipments currently move by rail; the rest is hauled by barge or truck. Each of these transportation systems faces problems that must be overcome before more coal can be moved. Most of the solutions come down to money. William H. Demsey, president of the Association of American Railroads, told the House Ways and Means Committee May 26, 1977, that the railroads would have to acquire between 9,700 and 13,400 new coal cars each year to handle the increase in coal traffic.

Demsey told the committee that the railroads would have little difficulty coming up with the money for new cars and locomotives. Where the railroads face severe problems, Demsey said, is in their deteriorating roadbeds, particularly "mainlines in the Northeast, some areas of the Midwest and secondary rail lines throughout the nation...." A recent study by the Department of Transportation concluded that $5 billion would have to be spent to upgrade rails and run new lines needed to haul coal in 1985.[12] The General Accounting Office concluded that roadbed repairs can be made, "but federal action may be needed. The railroads have the capability to expand, but expansion will not be without problems, particularly capital acquisition."[13]

The Department of Transportation study estimated that about $6 billion would have to be spent to upgrade highways in Appalachia, where at least half the nation's coal will be produced by 1985. The report said that "the projected increase in coal production will place unprecedented demands upon the trucks and highway systems, especially in the East.... Ap-

[11] Edmund Faltermayer, "What the Coal Strike Has Obscured," *Fortune,"* April 10, 1978, pp. 30-31.

[12] U.S. Department of Transportation, "Transporting the Nation's Coal — A Preliminary Assessment," January 1978, p. II-19.

[13] "U.S. Coal Development — Promises and Uncertainties," *op. cit.*, p. 5.1. See also "Future of American Railroads, *E.R.R.*, 1978 Vol. I, pp. 181-200.

palachia's coal road problems could well become so severe as to become a bottlneck on coal production."[14]

Slightly more than 10 per cent of all coal shipments today are hauled via the nation's inland waterways, primarily the Ohio and Mississippi river systems. The Department of Transportation report concluded that some locks in the inland water system would "hit their capacity in the relatively near future" and this, too, could be a bottleneck in coal production. The study predicted that by 1985 about 257.5 million tons of coal — or more than 20 per cent of total coal production — will move on the inland waterways.

Bidding to compete with the railroads' share of the coal-hauling market are several companies proposing to pulverize coal into fine particles and mix it with fluid into a "slurry," which would be pumped through pipelines from mine to user. Coal slurry pipelines have their own problems, however. A slurry pipeline requires enormous amounts of water at the point of shipment — a key problem in the arid western coal fields. There is also the problem of disposing of pipeline water that has been contaminated by the coal it carries in suspension. The Department of Transportation concluded that pipelines will play "only a minor role in the transportation of coal through 1985" and questioned their feasibility for the long-term future.[15]

Impact of Coal Use on the Environment

Environmental problems could pose the greatest deterrent to increased coal production. "The environmental issue is paramount," the General Accounting Office concluded. "We cannot use one billion tons of coal in one year without harming our environment. At least not with current technology."

Under President Carter's proposed energy plan, all new industrial and utility plants must employ the best available pollution control technology. The technology currently considered to be the best available is flue gas desulfurization devices or "scrubbers." The scrubbers remove most of the sulfur from the smoke and gasses that escape up the chimney and into the air. They are expensive to install and run and they produce a limestone sludge that is itself a disposal problem. The General Accounting Office estimates that the capital costs of controlling emissions under the Carter plan will be $19.1 billion by 1985, with annual operating costs of $1.3 billion.

The administration believes industry is overstating the problem of meeting air pollution requirements. "The en-

---

[14] "Transporting the Nation's Coal — A Preliminary Assessment," *op. cit.*, p. II-12.

[15] Congress is considering legislation (HR 1609) that would make it easier for pipeline companies to build their lines. They would be granted the right of eminent domain — to force private owners to sell their property.

vironmental constraints are no problem for 80 per cent of the country, with 85 per cent of the population," John O'Leary said. "There are some areas where you won't be able to site a coal-fired plant, such as in the Los Angeles Basin or perhaps around Houston, but that leaves lots of room."

O'Leary told the annual meeting of the National Coal Association last June that public health is ultimately at stake in the pollution debate. "It's not aesthetics, it's purely public health," he said. "The best available control technology orientation is sensible and prudent and indeed...it is the only acceptable course open to the country."[16]

Many environmentalists feel that shifting to a coal-based energy program, even using the best available pollution control technology, is a mistake. The environmental problems associated with coal production include devastation of the land through strip mining, water pollution, air pollution and an increase in respiratory diseases.[17] "Carter's promise that coal conversion could be achieved 'without endangering the public health or degrading the environment' cannot be taken at face value," said environmental writer Helen Sandalls. "A coal conversion program simply cannot have public health as its primary concern."[18]

The administration acknowledges that environmental problems, particularly increased air pollution, will result from switching to a coal economy. But according to John O'Leary, the problem can be controlled for the next seven or eight years because the nation already is reducing air pollution. The reductions should be large enough to keep the added burden caused by coal burning from bringing the totals over existing levels. But that margin will not be there if coal consumption is increased beyond the projected 1985 levels, O'Leary said.

Last October a committee headed by Dr. David Rall, director of the National Institute of Environmental Health Sciences, looked at the health and environmental effects of burning coal. The committee concluded that "it is safe to proceed with [Carter's energy plan] through 1985 if strong environmental and safety policies are followed.... Even with the best mitigation policies, there will be some adverse health and environmental effects for the dramatic increase in coal use. However, these will not impact all regions and individuals uniformly."[19]

[16] National Coal Association, "Proceedings from 60th Anniversary Convention," June 26-28, 1977, p. 19.

[17] According to the American Lung Association, sulfur oxides in the air aggravate asthmatic conditions, heart and lung disease in the elderly, respiratory disorders in children, and emphysema, and increase the risk of lung cancer.

[18] Helen Sandalls, "Coal and Consequences," *Environmental Action*, May 21, 1977, p. 3.

[19] "Health and Environmental Effects of Increased Use of Coal Utilization," *Federal Register*, Vol. 43, No. 10, Jan. 16, 1978, p. 2230.

Rise and Fall of King Coal

N ITS EARLY DAYS, coal was a simple oddity, a rock that burned. But in the burning, the black rocks released noxious odors and foul gases. Coal was such a smelly source of heat that England's King Edward I (1239-1307) ordered the death penalty for those found guilty of burning it. But within a few centuries the regal objection was overridden by economics.

Coal began its rise to dominance in the 16th century. English brick manufacturers found that coal was a useful fuel for their ovens. A serendipitous benefit was that bricks fired in coal-burning ovens were more fire resistant. This allowed people to burn coal on their hearths without jeopardizing their houses, leading to greater use of coal in home heating. Ralph Waldo Emerson, in his 1860 essay "Conduct of Life," described coal as "a portable climate."

The development of the iron market in England in the 17th and 18th centuries opened vast markets for coal, but development in the United States was slower. Coal did not come into widespread industrial use in the United States until the middle of the 19th century. The American iron industry developed slowly, using wood instead of coal to fuel the process that turns ore into metal. The iron industry of early America has been described as "more adept at obtaining tariff protection against English iron than in improving methods of production."[20]

The coal industry expanded enormously in the late 1800s. The first commercial shipments from "the Pittsburgh seam," the richest coal bed in the nation, came in 1853, just six years before the discovery of oil in western Pennsylvania. The invention of the steam locomotive gave further impetus to coal-production — both as a consumer of coal to fire its boilers and as a means of transportation that opened vast new markets. By the 1920s, coal was king in the United States. In 1925, 80 per cent of all energy in the country came from what English folk songs called "those dusty diamonds."

John L. Lewis and Early Miner Militancy

No one individual dominated the coal industry as John D. Rockefeller dominated the oil industry in the late 19th and early 20th centuries. One individual, however, did dominate the labor history of the coal fields. He was John L. Lewis. As president of the United Mine Workers of America from 1920 to 1960, Lewis controlled one of the most powerful and militant unions in the

[20] S.E. Morrison, H.S. Commager, W.E. Leuchtenburg, *The Growth of the American Republic*, Vol. I (1969), p. 455.

United States. Throughout his 40 years of leadership, the ability of the UMW to influence the country through strikes, slow-downs and threats of violence was legendary.

The roots of miner militancy are found in the work. Mining coal is dirty and dangerous. For most of its history, the work was pick and shovel stuff, demanding little skill but much muscle. In addition, early coal operators frequently took advantage of their workers. The relationship between miners and mine owners in the early 1900s was described by Thomas N. Bethell, formerly UMW research director and now editor of *Coal Patrol* newsletter.

> The coal operators were unimaginative in most things [Bethell wrote], but they were resourceful when it came to gouging the miner. They charged him for his tools, they charged him for the explosives he used to get out their coal; they charged him for his home; they charged him for trading at the company store and paid him in scrip to keep him from trading anywhere else; they charged him for the company doctor; they charged him for sending his kids to the company school; they provided no pension, but they took a paycheck deduction to cover the cost of burying him.[21]

Kentucky lawyer Harry Caudill, in his classic book on coal mining in Appalachia, *Night Comes to the Cumberlands,* described what happened to a miner who lost a leg in an accident in 1919: "This man still remembered the kindness of the Big Boss. On the day after he left the hospital with his crutches and new cork leg, he was told by the superintendent that he could remain in the house for a whole month, rent free, as a donation from the company, 'but be sure to get out by the end of the month.' "[22]

When Lewis became UMW president in 1920 he took over the largest union in the American Federation of Labor. But he faced immediate troubles. The end of World War I meant problems for the coal industry, which had geared up to produce for the war and now faced overproduction. The coal economy declined, and the depression that arrived in the coal fields starting in 1922 helped the operators to break the union.

In 1919 the UMW controlled 70 per cent of the nation's coal production, but by 1929 that figure had slipped to 40 per cent, and was as low as 10 per cent in West Virginia, Kentucky and Tennessee. Membership in the UMW dropped from 500,000 in 1923 to 150,000 in 1932.

The Roosevelt administration's New Deal gave the UMW a

---

[21] Thomas N. Bethell, "The UMW: Now More Than Ever," *The Washington Monthly,* March 1978, pp. 16-17.

[22] Harry M. Caudill, *Night Comes to the Cumberlands* (1962), p. 121.

Injuries in U.S. Bituminous Coal Mines

(Rate per million work hours in parentheses)

Year	Fatal	Disabling
1971	176 (0.72)	11,363 (46.41)
1972	153 (0.59)	12,012 (45.96)
1973	131 (0.48)	10,880 (40.06)
1974	131 (0.45)	8,298 (28.45)
1975	154 (0.42)	10,856 (29.89)
1976	140 (0.37)	14,096 (37.29)
1977*	128 (0.34)	14,512 (37.21)

*Preliminary and subject to change

Source: Mine Enforcement and Safety Administration, Department of Labor

new lease on life. The National Industrial Recovery Act of 1933 and the Wagner Act of 1935 protected the right of workers to join unions. Lewis geared up a massive organizing drive, struck the bituminous fields in 1933, and was successful. By 1940 UMW membership had grown to 600,000, "a rate of growth never seen before or since in the American labor movement."[23]

During the 1940s Lewis won an impressive series of victories for the union. His efforts were aided by the miners' willingness to strike at a moment's notice, to stay out far longer than anyone could predict and to defy the law. By 1950 coal miners were the highest paid industrial workers in the United States, with a health and pension plan that was the envy of all organized labor. But Lewis achieved those ends through such naked applications of power that he was hated and feared outside his union.

The most recent biography of Lewis concludes: "Lewis's life and career expose the inseparable relationship between means and ends. Himself a practitioner of the theory that power is the only morality, Lewis used every instrument at his command to accumulate power. Brutality, bullying, deceit, and bluff were all means Lewis used to achieve his ends."[24]

Switch to Oil, Gas After Second World War

The 1950s saw the decline of coal as the nation's leading energy source and, with that, the transformation of John L. Lewis from union tyrant into labor statesman. The decline was caused by oil and gas — energy sources that were then considered cheap, clean, and not subject to being cut off by militant

---

[23] Bethell, *op. cit.*, p. 18.

[24] Melvin Dubofsky and Warren Van Tine, *John L. Lewis* (1977), p. xv.

workers. The nation's industries and homeowners were discovering that these fuels had few of coal's inherent limitations: "Coal is dirty; it is bulky; it seldom occurs where you need it; and it varies widely in quality, in terms of chemical impurities, heat content, and combustion characteristics."[25]

By 1950 coal was providing only 34 per cent of the total energy demand in the United States. By 1975 it had declined to 14 per cent. Commercial and household uses of coal practically disappeared. Industrial use declined from 46 per cent in 1950 to 19.5 per cent in 1975. Only electric utilities continued to use coal as they had in the past. In 1950, 45 per cent of the energy consumed by electric utilities came from coal; in 1975 it was 44 per cent.[26]

From the miners' standpoint, this transformation meant fewer jobs. An increase in mechanization in the mines made the job situation even worse. The great membership gains the UMW experienced in the 1930s and 1940s evaporated. By 1955, only 200,000 workers were members of the UMW, a figure that was to fall even farther during the 1960s.

The UMW's response to these losses was not to fight to protect the jobs of all miners, but to accommodate. Lewis reasoned that it was better to have 200,000 well-paid miners than to have 400,000 on strike earning nothing. In exchange for progress on pay and benefits, Lewis agreed with the mine operators to forbid work disruptions and to eschew militancy. He fundamentally changed the adversary relationship that had existed in the coal fields by forging an allinace with the operators. Lewis even went so far as to give up the cherished mine workers' principle of "no contract, no work," permitting the 1952 contract to remain in force until 1955, requesting a new agreement only when he felt coal's financial prospects had improved.

Recent Unrest in the Mine Workers' Union

When Lewis retired in 1960, he left a union vastly diminished in size and power, in an industry that was equally hard-pressed. The union quickly became enmeshed in bitter internal struggles. The state to which the union had declined was shown by one chilling incident. On December 30, 1969, just six months after John L. Lewis died, Joseph A. "Jock" Yablonski, his wife, and his daughter were murdered by gunmen hired by W. A. "Tony" Boyle, then UMW president. Yablonski had been a Lewis lieutenant for many years and was challenging Boyle's election in the courts. Boyle, also a Lewis lieutenant for many

---

[25] "U.S.Coal Development-Promises, Uncertainties," *op. cit.*, p. 1.1.
[26] *Ibid.*, p. 1.2.

years, was arrested, tried, and, in 1974, convicted of Yablonski's murder.

The current UMW president, Arnold Miller, was elected in December 1972 as the miners ousted Boyle on a wave of reform sentiment. Miller's election marked a resurgence of the militancy that had been dormant for many years. But by most estimates Miller has been a weak leader. During the 1977-78 strike he was unable to control either the union's bargaining council or union members.

Miller's tenure may end soon. He recently suffered a stroke and a mild heart attack and his doctors are reportedly advising him to resign. And the miners are restive. Thomas N. Bethell recently described the union as "a half-sunken hulk with a certifiably incompetent captain sill roaming the bridge, eyes glassy with fear and fatigue, while the roar of mutiny in the engine room grows in volume." Bethell predicted that "within months, a new wave of wildcat strikes is almost certain to break across the coal fields as miners rebel against the new contract, and that could put the ship under once and for all."[27]

Long-Term Outlook for Coal

SHOULD AMERICA revert to a coal-based energy program, significant changes in the environment and in public health and safety could occur. These changes are likely to produce impetus for changes in the law and for research and development into ways to make coal a more acceptable source of power.

Current air pollution laws regulate some of the major pollutants that result from burning coal — sulfur oxides and particulates larger than a micron.[28] Some currently unregulated pollutants may be quite harmful, making them targets for regulation in the future. One category of unregulated pollutants that will likely increase as a result of increased coal use is submicron-sized particulates. These may pose a special public health hazard because they penetrate the respiratory system's natural filters and lodge deep within the lungs.

Trace elements form another category of unregulated pollutants. Coal gases contain small but significant quantities of mercury, lead, uranium, beryllium, arnsenic, fluorine, cadmium and selenium. "Data about them are limited," the

---

[27] Thomas N. Bethell, "UMW in the Pits," *The New Republic*, April 1, 1978, p. 9.
[28] A micron is a unit of length equal to one thousandth of a millimeter.

General Accounting Office reported, "but enough is known to suggest that they could cause serious consequences."[29]

Concern About High Carbon Dioxide Levels

Another environmental issue that will likely receive more attention in the future is long-term climate change. Burning coal increases the amount of carbon dioxide in the atmosphere. The Rall committee, formed to advise the President on the health and environmental effects of coal burning *(see p. 190),* said that the concentration of carbon dioxide in the atmosphere has increased steadily since 1860 and that it "could reach two or three times its present value within the next 100 years."[30]

For some time scientists have been concerned about carbon dioxide and the "greenhouse effect," a global warming trend. The carbon dioxide in the atmosphere permits sunlight to penetrate through it but prevents energy in the form of heat from radiating back into space. As a result, the overall climate could get warmer if carbon dioxide builds up. That in turn could lead to melting of the polar ice caps, flooding along most coastlines, changes in rainfall patterns and widespread disruption of agriculture as arid areas get rain and wet areas experience drought. These prospects led George M. Woodwell, director of the ecosystems center at the Woods Hole (Mass.) Marine Biological Laboratory, to warn: "Carbon dioxide, until now an apparently innocuous trace gas in the atmosphere, may be moving rapidly toward a central role as a major threat to the present world order."[31]

President Carter has requested nearly $3 million to study the long-term effects on the atmosphere of carbon dioxide build-up from burning coal and other hydrocarbons. Yet another by-product of burning coal is the phenomenon of "acid rain." Not much is known about what dangers acid rain might pose, but it is known that sulfur and nitrous oxides can create it, and that burning coal creates them. According to Dr. James M. Galloway of the University of Virginia, more than 2,000 Scandinavian lakes have been contaminated by acid rain, as have between 100 and 200 lakes in the New York State Adirondack Mountain region. Fish cannot reproduce in those lakes, Dr. Galloway said.[32]

The prospect of a coal-based energy system should also create a push for research and development into ways to make

---

[29] "U.S. Coal Development—Promises, Uncertainties," *op. cit.,* p. 9.8.

[30] "Health and Environmental Effects of Increased Use of Coal Utilization," *op. cit.,* p. 2231.

[31] George M. Woodwell, "The Carbon Dioxide Question," *Scientific American,* January 1978, p. 43. See also "World Weather Trends," *E.R.R.,* 1974 Vol. II, pp. 517-538.

[32] Quoted in *Congressional Quarterly Weekly Report,* June 18, 1977, p. 1218.

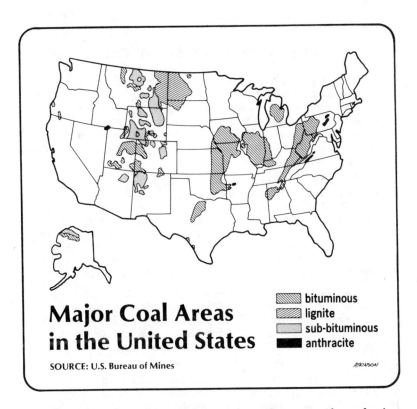

Major Coal Areas in the United States

bituminous
lignite
sub-bituminous
anthracite

SOURCE: U.S. Bureau of Mines

ATKINSON

coal a cleaner fuel. Several research avenues currently under investigation are likely to accelerate as coal use increases. Some are already very close to commercial application.

Gasification, Liquefaction and Methanation

Coal gasification — changing coal into a gas that can be burned cleanly — is quite close to everyday use. The Southern California Edison Co., an electric utility, is reported to be ready to announce a gasification project very soon. The project will convert western low-sulfur coal into a gas that can be burned to generate electrical power. Southern California Edison is a leader in this field because it is fairly close to the vast western coal fields *(see map)* but faces heavy environmental problems in burning coal in the smoggy Los Angeles area.

To convert coal to synthetic gas, the coal is fed with steam and oxygen into a high temperature pressurized reactor. The result is a product called low-Btu gas. A Btu is a British thermal unit, the amount of energy necessary to raise the temperature of one pound of water one degree Fahrenheit. Low-Btu gas has a lower heat value than natural gas, but could be a valuable boiler fuel. A process called methanation — the chemical reaction of carbon monoxide and hydrogen to produce methane and water — can be used to convert low-Btu gas into high-Btu gas. High-

197

Western Coal — A Special Case

Coal mining in the western states has become one of the most controversial aspects of the whole issue of the development of a coal economy. The West possesses vast deposits of coal, technically called sub-bituminous. It is of lower heat value than the coal in the East, but it contains less sulfur, which means it burns cleaner than eastern coal. In addition, it can be easily strip-mined.

Strip mining, already under way at the time of the 1973-74 Arab oil embargo, boomed in its aftermath. America, in its quest for self-sufficiency in fuels, looked westward. The Interior Department reports that the western states currently produce 56 million tons of coal a year, most of it strip-mined. Under the Ford administration's Project Independence, that figure was projected to increase to 250 million tons by 1980 and 340 million tons by 1985. Under the Carter plan, the figure is 250 million by 1985.

Talk of further — and accelerated — coal development has created much disquiet in the West, not only among traditional environmentalists but also among ranchers who see the despoliation of grazing land. A 1977 federal law requires reclamation, but there is doubt that strip-mined land can be effectively reclaimed for grazing in the arid West.

The West's scarcity of water relates to strip mining in another way. Western coal is far from market, making transportation expensive. To solve that problem, electric utilities have developed large generating plants at the mouth of strip mines, so they can burn the coal as it emerges from the ground. But these plants require large amounts of water, as do other alternatives such as coal slurry pipelines and coal gasification and liquefaction.

Finally, there is the emergence of western boom towns — with all of their problems — as a consequence of the strip mining boom. The sociological term "Gillette syndrome" is used to describe what vast infusions of cash and transient workers can do to a small, essentially rural community. The population of Gillette, Wyo., has more than tripled in 15 years to over 11,000. Most of the increase has come in the last five years. Such other boom towns as Craig, Colo., and Colstrip and Rock Springs, Wyo., are similarly beset by high rates of crime, alcoholism and divorce, overcrowded schools and a dearth of social services, including medical care.

Btu gas has approximately the same heat content as natural gas, making it a complete substitute.

Researchers also are looking at ways to turn coal into a liquid — a synthetic oil. Products created by coal liquefaction could be used as substitutes for oil in boiler fuels and as substitutes for gasoline, heating oil, diesel oil and petroleum feedstocks in the petrochemical industry. Currently the Department of Energy is supporting four research projects related to liquefaction, and the major problems seem to be technical ones, such as the durability of equipment and development of the proper catalysts. The Department of Energy estimates that the United States will see the commercial application of liquid coal by 1990.

A third technological approach may be the closest to large-scale commercial application. This approach is called magnetohydrodynamics (MHD), a method of burning coal that results in direct electrical generation, without the need for boilers and turbines. "Twenty years of research indicates that MHD will substantially increase the overall efficiency of the process of converting the energy in coal into electric energy; that it will mean a giant step forward in controlling sulfur emissions; and that it will need significantly less water than conventional coal-burning or nuclear power plants."[33]

The Soviet Union is considerably ahead of the United States in practical applications of magnetohydrodynamics. It already has a power plant functioning and expects to have a 400 megawatt commercial plant in operation by the 1980s. The Soviet Union, like the United States, has enormous coal deposits.

If the technological approaches to curing coal's ills are successful, it appears that coal could have another long reign in the United States. But it is important to remember that coal's problems are more than technical. They are human as well, as the 160,000 members of the United Mine Workers recently demonstrated. So coal's future remains obscured by the smoke of labor warfare, dirtied by environmental issues, and slowed by a troubled transportation system. The National Coal Association continues to assert that "coal is America's ace in the hole." But it is also possible that coal could turn out to be the rotten apple in the nation's energy barrel.

---

[33] Joan Melcher, "Cleaner Coal Conversion," *Environment*, March 1978, p. 12.

Selected Bibliography

Books

Caudill, Harry M., *Night Comes to the Cumberlands*, Little, Brown, 1962.
——*The Watches of the Night*, Little, Brown, 1976.
Dubofsky, Melvyn and Warren Van Tine, *John L. Lewis*, Quadrangle, 1977.
Freeman, S. David, *Energy: The New Era*, Walker, 1974.
Park, Charles F. Jr., *Earthbound — Minerals, Energy, and Man's Future*, Freeman, Cooper, 1975.

Articles

Bethel, Thomas N., "The UMW: Now More Than Ever," *The Washington Monthly*, March 1978.
Cook, James, "Do Coal and Oil Mix?" *Forbes*, Oct. 15, 1977.
Faltermayer, Edmund, "What the Coal Strike Has Obscured," *Fortune*, April 10, 1978.
Kirschten, Dick, "Watch Out! The Great Coal Rush Has Started," *National Journal*, Oct. 29, 1977.
——"Converting to Coal—Can It Be Done Cleanly," *National Journal*, May 21, 1977.
Melcher, Joan, "Cleaner Coal Conversion," *Environment*, March 1978, pp. 12-17.
Rankin, Bob, "Coal Conversion: Key to Carter Plan," *Congressional Quarterly Weekly Report*, June 18, 1977.
Sandalls, Helen, "Coal — and Consequences," *Environmental Action*, May 21, 1977.
Sheets, Kenneth R., "Some Second Thoughts About Coal," *U.S. News & World Report*, March 13, 1978.
Walsh, John, "Texas Is Testing Ground for Impact of Coal Use on Economic Growth," *Science*, Nov. 11, 1977.
Woodwell, George M., "The Carbon Dioxide Question," *Scientific American*, January 1978.

Reports and Studies

American Association for the Advancement of Science, "Energy, Water, and the West," 1976.
Ford Foundation, Energy Policy Project, "A Time to Choose," 1974.
General Accounting Office of the United States, "U.S. Coal Development — Promises, Uncertainties," Sept. 22, 1977.
Department of Health, Education and Welfare, "Health and Environmental Effects of Increased Use of Coal Utilization," Jan. 16, 1978.
Editorial Research Reports, "Coal Negotiations," 1974 Vol. II, p. 807; "Future of American Railroads," 1978 Vol. I, p. 181; "Oil Antitrust Action," 1978 Vol. I, p. 101; "Strip Mining," 1973 Vol. II, p. 861.
Department of Transportation, "Transporting the Nation's Coal — A Preliminary Assessment," January 1978.
National Coal Association, "Proceedings from 60th Anniversary Convention National Coal Association Washington, D.C., June 26-28, 1977."

INDEX